INTERNATIONAL CENTRE FOR MECHANICAL SCIENCES

ZLATKO JANKOVIĆ
UNIVERSITY OF ZAGREB

A CONTRIBUTION TO THE VECTOR AND TENSOR ANALYSIS

COURSE HELD AT THE DEPARTMENT
FOR MECHANICS OF DEFORMABLE BODIES
SEPTEMBER - OCTOBER 1969

Springer-Verlag Wien GmbH 1969

COURSES AND LECTURES - N. 6

ISBN 978-3-211-81151-1 ISBN 978-3-7091-2888-6 (eBook)
DOI 10.1007/978-3-7091-2888-6

These lecture notes are manuscripts of three papers which will be published in TENSOR, vol. 21 (1970).

P R E F A C E

I was very much honoured by Professor Sobrero's kind invitation to deliver a series of lectures on vector and tensor calculus at the Centre for Mechanical Sciences. The course was intended to review the existing state and to report my investigations in this field of research. Since the main part of the series (18 lectures) was devoted to this aim, it seemed to me that the best way of fulfilling the task of the lecture notes would be to copy the manuscripts of my three papers (to be published in Tensor 21-1970), which are concerned with the above problems.

The vector and tensor calculus is an important branch of mathematics and a very efficient tool in many fields of natural sciences and their applications. So, it seemed worth attempting to promote its present state by a more general formulation which, at the same time, would be a simple and plausible scheme to cover, besides others, all applications in classical, relativistic and quantum mechanics.

The most characteristic feature of the present scheme is the direct approach to vectors and tensors which, by using an appropriate notation, resulted into general, compact and straightforward rules. In order to emphasize the logical sequence of the most important steps for building up the scheme, only the most impor-

tant topics have been analyzed in full detail, and the introduction of each paper contains the respective summary. I shall be happy if some of the participants to the lectures or readers of these notes will find some interest to elaborate the remaining aspects.

At the end I would like to express my deepest gratitude and appreciation for the opportunity of giving these lectures in such an important institution as the Centre for Mechanical Sciences, among such outstanding scientists, dear colleagues and before such a selected audiance. In particular, we are all very much indebted to Professor Luigi Sobrero, the Secretary General of the Centre and Director of the Institute for Mechanics of the University of Trieste, for the foundation of the Centre and for the very successful promotion of the efficient working atmosphere in this basic institution in the field of modern mechanical sciences.

Z.Janković

Zagreb, January 1970

A CONTRIBUTION TO THE VECTOR AND TENSOR ALGEBRA

The aim of this paper is to give a simple and suitable approach to the vector and tensor algebra which represents a general and natural scheme for a wide range of applications. The essential feature of this scheme is the synthesis of notions of contravariance and covariance with the bra and ket forms. The main consequences of such an approach are investigated and an appropriate notation is developed.

In Chapter I we consider the basic concepts. Chapter II contains the discussion of the connection between the contravariant and covariant forms, while in Chapter III the relations between bra and ket vector forms are investigated. In Chapter IV the transformation properties under the change of basis vectors are established. Chapter V contains an extension and application of the obtained results to tensors. In Chapter VI some remarks are added.

CHAPTER I

The basic assumptions

We briefly recall [1] that a vector (linear) space is composed of an additive Abelian group X (whose elements are called vectors), a scalar body K and a multiplication, so that for each ordered pair $(\alpha, \mathbf{a})$ $(\alpha \in K, \mathbf{a} \in X)$ there exists one vector $(\alpha \mathbf{a}) \in X$ in such a way that the following axioms are fulfilled

$$\mathbf{a} + \mathbf{b} = \mathbf{b} + \mathbf{a} ,$$

$$(\mathbf{a} + \mathbf{b}) + \mathbf{c} = \mathbf{a} + (\mathbf{b} + \mathbf{c}) ,$$

$$(\alpha \beta) \mathbf{a} = \alpha (\beta \mathbf{a}) , \qquad (1.1)$$

$$\alpha (\mathbf{a} + \mathbf{b}) = \alpha \mathbf{a} + \alpha \mathbf{b} ,$$

$$(\alpha + \beta) \mathbf{a} = \alpha \mathbf{a} + \beta \mathbf{a} ,$$

$$1 \cdot \mathbf{a} = \mathbf{a} .$$

Besides, the existence of the zero vector $\mathbf{0}$ and of the opposite vectors $-\mathbf{a}$ is postulated.

Since the dimension of the vector

space is the maximum number of linearly independent vectors in the vector space, we are able to choose n basis vectors in an n-dimensional vector space and represent each vector in a unique way as a sum of n terms; each term is a product of a basis vector and a scalar - the corresponding component of the vector.

Now, we take four vector spaces $X^{<}$, $X_{<}$, ${}^{>}X$, ${}_{>}X$, each with its system of basis vectors $e^{i}, e_{i}, {}^{i}e, {}_{i}e$ $(i = 1, 2, .., n)$, and call them "bra-up", "bra-down", "ket-up", "ket-down" vector spaces and basis vectors, respectively. In these vector spaces we have the vector representations

$$a^{<} = a_{i}\, e^{i} ; \quad a_{<} = a^{i}\, e_{i} ; \quad {}^{>}a = {}_{i}a\, {}^{i}e ; \quad {}_{>}a = {}^{i}a\, {}_{i}e , \tag{1.2}$$

$a_{i}, a^{i}, {}_{i}a, {}^{i}a$ are vector components (the same terms can be used for them as for the respective vector spaces) and belong to the same scalar body K . Here the usual convention upon the double index summation is adopted. In further development the terms "up" and "down" for vectors ("down" and "up" for components) will be related with the commonly applied terms "covariant" and "contravariant".

From the axioms (1.1) and the explicit forms (1. 2) it easily follows that in each of the vector spaces the sum (the difference) of vectors has as its components the sum (the difference) of the corresponding components of its members. The vector αa has as its components the components of the vector a multiplied by α , a representing any of the vectors (1. 2).

We define a multiplication between the bra and ket vectors, the so-called scalar product, which has to be a scalar quantity. The four possible combinations are symbolically written

$$a^{<}\,{}_{>}b \; , \quad a^{<}\,{}^{>}b \; , \quad a_{<}\,{}_{>}b \; , \quad a_{<}\,{}^{>}b \; . \qquad (1.3\,a)$$

For the scalar product we assume the following three properties :

First, we require that for each of the four scalar products (1. 3a), the left- and right-hand side distribution law be fulfilled

$$a\,(b_1 + b_2) = a b_1 + a b_2 \; ,$$

$$(a_1 + a_2)\, b = a_1 b + a_2 b \ . \tag{1.4}$$

Second, we assume that for all $\alpha, \beta \in K$ there exists

$$(\alpha a)\, b = \alpha\, a b\,, \quad a(\beta b) = \beta\, a b \ . \tag{1.5}$$

Third, we assume the following explicit values for the scalar products of basis vectors

$$e^{i}\,{}_{j}e = {}^{i}\delta_{j}\,, \qquad e_{i}\,{}^{j}e = {}_{i}\delta^{j}\,, \tag{1.6a}$$

$$e^{i}\,{}^{j}e = {}^{i}g^{j}\,, \qquad e_{i}\,{}_{j}e = {}_{i}g_{j}\ . \tag{1.6b}$$

The first two relations (1.6a) indicate that the corresponding basis vectors are "orthogonal". The symbol δ means the Kronecker symbol, while ${}^{i}g^{j}$ and ${}_{i}g_{j}$ in the second two relations (1.6b) are $2n^2$ scalar quantities, which should be given explicitly.

Because of the assumed properties (1.4), (1.5) and (1.6) the scalar products (1.3a) can be explicitly written in the forms

$$a^{<}\,{}_{>}b = a_{i}\,{}^{i}b\,, \qquad a_{<}\,{}^{>}b = a^{i}\,{}_{i}b\,,$$

$$a^{<\,>}b = a_i \, {}^i g^j \, {}_j b \,, \quad a_{<\,>} b = a^i \, {}_i g_j \, {}^j b \,. \qquad (1.3b)$$

The notation introduced here could be regarded as a "direct" system of notation [3]. We underline that the symbols „>" and „<" (or the left- and right-hand side indices of the basis vectors) represent the "active" side of a vector in the scalar multiplication, and are similar to "valences" which are saturated by the opposite valence of the other factor. At the same time Dirac's terms [2] "bra" and "ket" are explained to originate from the closed "bra-c-ket" "< >" in the scalar product.

The vector spaces, i.e. the basis vectors, the vector components (1.2), the ${}^i g^j$, ${}_i g_j$ (1.6b) may depend on some parameters x^k $(k = 1, 2, \ldots, m)$.

In this paper they are referred to one and the same set of parameter values, i.e. to the same point $P(x^k)$ of the m-dimensional parameter manifold. Consequently, we should indicate this fact by writing $X(P)$, $\mathbf{e}(P)$, $\mathbf{a}(P)$, $a_i(P)$, ${}^i g^j(P)$, ${}_i g_j(P)$ etc. We omit the sign P because all quantities in this paper refer to one and the same point $P(x^k)$ of the parameter manifold, no connection being assumed here between the vector spaces referring to different points of the parameter manifold.

CHAPTER II

The connection between the bra(ket)-up and - down vector spaces

Now, we introduce the connecting quantities (i. e. linear operators) between the vector spaces. First, we define the "identity operators"

$$ {}_{>}E^{<} = {}_{i}e\ {}^{i}\delta_{j}\ e^{j} = {}_{i}e\ e^{i}\ ,\quad {}^{>}E_{<} = {}^{i}e\ {}_{i}\delta^{j}\ e_{j} = {}^{i}e\ e_{i}\ , \qquad (2.1)$$

for the vector spaces $X^{<}$, ${}_{>}X$ and $X_{<}$, ${}^{>}X$, respectively. It is easy to find from (1.2), (1.4) and (1.6) that these operators act as identity operators in the mentioned spaces,

$$a^{<}\,{}_{>}E^{<} = a_j\, e^j\, {}_{i}e\, e^i = a_i\, e^i = a^{<} ,$$

$$a_{<}\,{}^{>}E_{<} = a_{<} , \quad {}_{>}E^{<}\,{}_{>}a = {}_{>}a , \quad {}^{>}E_{<}\,{}^{>}a = {}^{>}a , \qquad (2.2)$$

the basic property of the operators (2. 1) being

$$({}_{>}E^{<})^p = {}_{>}E^{<} , \quad ({}^{>}E_{<})^p = {}^{>}E_{<} , \qquad (2.3)$$

p a positive integer.

Further, we examine other two operators i. e. the "valence lowering operator"

$$ {}_{>}E^{<}\,{}^{>}E_{<} = {}_{i}e\, e^i\, {}^{j}e\, e_j = {}_{i}e\, g^{ij}\, e_j = {}_{>}g_{<} \qquad (2.4a)$$

and the "valence raising operator"

$$ {}^{>}E_{<}\,{}_{>}E^{<} = {}^{i}e\, e_i\, {}_{j}e\, e^j = {}^{i}e\, {}_{i}g_{j}\, e^j = {}^{>}g^{<} \qquad (2.4b)$$

We determine the effect of these operators

$$\bar{a}_{<} = a^{<}\,{}_{>}g_{<} , \quad \bar{\bar{a}}^{<} = \bar{a}_{<}\,{}^{>}g^{<} = a^{<}\,{}_{>}g_{<}\,{}^{>}g^{<} , \qquad (2.5)$$

$$\bar{a}^{<} = a_{<}\,{}^{>}g^{<} , \quad \bar{\bar{a}}_{<} = \bar{a}^{<}\,{}_{>}g_{<} = a_{<}\,{}^{>}g^{<}\,{}_{>}g_{<} ,$$

with two analogous relations for the ket vectors. The dashed vectors are vectors associated to the original ones in the other bra (ket) vector space. We put the natural requirement that the associated vector of the associated vector should be the original one

$$\bar{\bar{a}}^{<} = a^{<} , \quad \bar{\bar{a}}_{<} = a_{<} ,$$

$$ {}^{>}\bar{\bar{a}} = {}^{>}a , \quad {}_{>}\bar{\bar{a}} = {}_{>}a \; : \qquad (2.6)$$

To satisfy this requirement, we see from (2. 5) that the following conditions must be fulfilled

$$ {}_{>}g_{<}\;{}^{>}g^{<} = {}_{>}E^{<}\,, \quad {}^{>}g^{<}\;{}_{>}g_{<} = {}^{>}E_{<}\,, \tag{2.7a} $$

i. e. because of (2. 1) and (2. 4)

$$ {}^{i}g^{j}\;{}_{j}g_{k} = {}^{i}\delta_{k}\,, \quad {}_{i}g_{j}\;{}^{j}g^{k} = {}_{i}\delta^{k} \tag{2.7b} $$

or

$$ G\,g = E\,, \quad g\,G = E\,, \tag{2.7c} $$

where G and g are the matrices (${}^{i}g^{j}$) and (${}_{i}g_{j}$) (with non-vanishing determinants), respectively, E being the unit matrix. The conditions (2. 7) represent restrictions on the number of independent quantities (1. 6b). In fact, only n^2 quantities, for instance the components of the matrix g (or G), can be independent, the other n^2 following from them.

The relations (2. 7) express, indeed, the fact that the operators (2. 4) are inverse operators one to the other.

In the particular case

$$ {}_{>}g_{<} = {}_{>}E_{<}\,, \quad {}^{>}g^{<} = {}^{>}E^{<}\,; \text{ i.e. } {}^{i}g^{j} = {}^{i}\delta^{j}, \; {}_{i}g_{j} = {}_{i}\delta_{j} \tag{2.8} $$

the difference between the $X^<$ and $X_<$ (${}^>X$ and ${}_>X$) vector spaces disappears and they become identical.

Since the original and the associated vector (2. 5) are in one-to-one correspondence, in the general case we can identify the corresponding bra (ket)-up and -down vectors to be two forms of the same bra (ket) vector. In this way the number of vector spaces is reduced from four to two, one bra and one ket vector space, each of them with two forms of any vector.

With the help of the relations (1. 2), (1. 6), (2. 1), (2. 4), (2. 5) and (2. 7) we easily compose the table

${}_>a \equiv$	${}^>a \equiv$	$a^< \equiv$	$a_< \equiv$
${}_>E^<\ {}_>a$	${}^>E_<\ {}_>a$	$a_<\ {}_>E^<$	$a_<\ {}^>E_<$
${}_>g_<\ {}^>a$	${}^>g^<\ {}^>a$	$a^<\ {}^>g^<$	$a^<\ {}_>g_<$

(2. 9)

This table contains all connections between different forms of bra (ket) vectors, especially of basis vectors, as well as the transformation formulas for the vector components. We also compose an equivalent explicit table containing the results of the action of the identity operators (2.1)

	${}_{>}a \equiv$	${}^{>}a \equiv$	$a^{<} \equiv$	$a_{<} \equiv$
${}_{>}E^{<}$	${}^{i}a \; {}_{i}e$	${}^{ji}q \; {}_{i}a \; {}_{j}e$	$a_{i} \; e^{i}$	$a^{i} \; {}_{i}q_{j} \; e^{j}$
${}^{>}E_{<}$	${}_{j}q_{i} \; {}^{i}a \; {}^{j}e$	${}_{i}a \; {}^{i}e$	$a_{i} \; q^{ij} \; e_{j}$	$a^{i} \; e_{i}$

(2.10)

With the help of the relations (2.9) or (2.10) we find that all the four forms of the scalar products (1.3b) are identical

$$a^{<}{}_{>}b = a^{<}{}^{>}b = a_{<}{}_{>}b = a_{<}{}^{>}b = \mathbf{a}\,\mathbf{b} \,. \qquad (2.11)$$

CHAPTER III

The connection between the bra and ket vector spaces

We wish to establish simple connections between the bra and ket vectors, too. We therefore define the linear "transposition operators" operating between the bra and ket vector spaces

$$
\begin{array}{lll}
 & \text{Spaces} & \text{Operators} \\
A) & X_{<} \,,\; {}_{>}X & {}^{>}_{>}T \,,\; T^{<}_{<} \\
 & X^{<} \,,\; {}^{>}X & {}_{>}^{>}T \,,\; T^{<}_{<} \\
B) & X_{<} \,,\; {}^{>}X & {}^{>>}T \,,\; T_{<<} \\
 & X^{<} \,,\; {}_{>}X & {}_{>>}T \,,\; T^{<<}
\end{array}
\qquad (3.1)
$$

We explain the operation of these operators by the example of the first case $A)$

$$a_{<}\, {}^{>}_{>}T = {}_{>}\bar{a} \qquad , \qquad T^{<}_{<}\, {}_{>}\bar{a} = \bar{\bar{a}}_{<} \,. \qquad (3.2)$$

The dashed vectors are the vectors associated to the original ones in the other corresponding space. We put the natural requirement that the associated vector of the associated vector should be the original one

$$\bar{\bar{a}}_{<} = a_{<} \tag{3.3}$$

From the relations (3.2) and (3.3) we infer that the relation

$$T_{<}^{<} a_{<} {}_{>}^{>}T = a_{<} \tag{3.4}$$

has to be fulfilled identically. Because of

$$T_{<}^{<} = T^{i}_{j}\, e_i\, e^{j} \;, \quad T^{<}_{<} = T_{i}{}^{j}\, e^{i}\, e_{j} \;, \qquad etc. \tag{3.5}$$

from the explicit form of (3.4) one finds that the components of the transposition operators (3.5) should satisfy the relation

$$T^{i}{}_{j}\; {}_{r}{}^{j}T = {}^{i}\delta_{r} \tag{3.6}$$

In an analogous manner we can discuss the other cases A) and B).

The most simple transposition operators, satisfying the conditions such as (3.6) are

$$\text{A)}\quad X_{<}\,,\ {}_{>}X \qquad {}^{>}{}_{>}E = {}^{i}e\,{}_{i}e\,, \quad E_{<}{}^{<} = e_i\, e^i\,,$$

$$X^{<}\,,\ {}^{>}X \qquad {}_{>}{}^{>}E = {}_{i}e\,{}^{i}e\,, \quad E^{<}{}_{<} = e^i\, e_i\,.$$

$$\tag{3.7}$$

$$\text{B)}\quad X_{<}\,,\ {}^{>}X \qquad {}^{>>}E = {}^{i}e\,{}^{i}e\,, \quad E_{<<} = e_i\, e_i\,,$$

$$X^{<}\,,\ {}_{>}X \qquad {}_{>>}E = {}_{i}e\,{}_{i}e\,, \quad E^{<<} = e^i\, e^i\,.$$

For the choice of transposition operators (3.7) we find the corresponding basis vectors and the corresponding vector components applying the relations analogous to (3.2)

	e^i	e_i	${}^{i}e$	${}_{i}e$
A)	${}^{i}e$	${}_{i}e$	e^i	e_i
B)	${}_{i}e$	${}^{i}e$	e_i	e^i

(3.8)

a)

	$\bar{a}_i$	$\bar{a}^i$	${}_{i}\bar{a}$	${}^{i}\bar{a}$
A)	${}_{i}a$	${}^{i}a$	a_i	a^i
B)	${}^{i}a$	${}_{i}a$	a^i	a_i

(3.9a)

We could also apply the transposition operators ${}_{>}\overset{*>}{E}$, $E_{<}^{<*}$ etc. of the same form as those in (3.7) but which change the vector components, belonging to the valences saturated by a valence with an asterisk, to their complex conjugate values. We illustrate this by the example (3.2)

$$ {}_{>}\bar{a} = a_{<}\,{}_{>}\overset{*>}{E} = (a^{i}\,e_{i})\,{}_{>}\overset{*>}{E} = a^{i*}\,{}_{i}e \qquad (3.10)$$

the relation (3.3) being automatically satisfied. Since the effect of the operators ${}_{>}\overset{*>}{E}$, $E_{<}^{<*}$ etc. on the basis vectors is the same as the effect of the operators in (3.7), the connections (3.8) are also valid for them. However, the components of the associated vector and the components of the original vector are connected in the following way

$b)$	$\bar{a}_{i}$	$\bar{a}^{i}$	${}_{i}\bar{a}$	${}^{i}\bar{a}$	
A)	${}_{i}a^{*}$	${}^{i}a^{*}$	a_{i}^{*}	a^{i*}	(3.9b)
B)	${}^{i}a^{*}$	${}_{i}a^{*}$	a^{i*}	a_{i}^{*}	

In both cases $a)$ and $b)$ it is also possible to change the sign.

A natural consequence of our discussion is to identify the associated vector with the representation of the original vector in the respective vector space, thus connecting the bra and ket forms of the same vector in one-to-one way. This connection, together with the connection of the bra (ket)-up and -down forms of a vector shows that each vector can be represented in four equivalent forms in four vector spaces. Briefly, we could say that the same vector can be represented in one vector space X in four different but equivalent forms, for each of these forms the axioms (1. 1) being valid separately. The connection between the bra and ket-up and -down forms are established by (2. 9) and (3. 8), (3. 9) respectively, while the scalar product of two vectors between their bra and ket forms is defined by (1. 4), (1. 5), (1. 6) and (2. 11).

The consequences of the connections A, B, a, b, regarding the valence lowering and raising operators (2. 4) and the scalar product (2. 11), follow from the relations (2. 5), (2. 7), (3. 8) and (3. 9). They are contained in the table

	$g =$	$G =$	$a_{<}{}^{>}b = a^i\ {}_i b =$	
Aa)	$\tilde{g}$	$\tilde{G}$	$b^{<}{}_{>}a = b_i\ a^i$	
Ab)	$g^{\dagger}$	$G^{\dagger}$	$(b^{<}{}_{>}a)^* = b_i^*\ a^i$	(3.11)
Ba)	$\tilde{G}$	$\tilde{g}$	$b_{<}{}^{>}a = b^i\ a^i$	
Bb)	$G^{\dagger}$	$g^{\dagger}$	$(b_{<}{}^{>}a)^* = b^{i*}\ a^i$	

The symbols have the usual meaning : $\sim$ the transposed matrix, $\dagger$ the Hermitian conjugate matrix, $*$ the conjugate complex quantity.

The matrix g (and G) can be represented in two more specific ways

$$g = g_1 + i g_2 = g_1^S + i g_2^S + g_1^A + i g_2^A = g^S + g^A \quad , \qquad (3.12)$$

where g_1 and $i g_2$ mean the real and imaginary matrices, g^S and g^A being the symmetric and antisymmetric matrices, respectively

$$g^S = \frac{1}{2}\ (g + \tilde{g}) \quad ,$$

$$g^A = \frac{1}{2}(g - \tilde{g}) \tag{3.13}$$

Substituting (3.12) into (2.7c), we obtain

$$(g_1 G_1 - g_2 G_2) + i(g_1 G_2 + g_2 G_1) = E \quad , \tag{3.14}$$

and an analogous relation by interchanging g and G in (3.14). Hence we determine

$$g_1 = [G_1 + G_2 G_1^{-1} G_2] \quad , \quad g_2 = -[G_2 + G_1 G_2^{-1} G_1] , \tag{3.15}$$

and two analogous relations by interchanging g and G in (3.15).

The relations (3.15) show again that the g and G matrices determine each other completely, so that in the case of complex matrices there are at most $2n^2$ quantities at our disposal (for instance the real and imaginary parts of components of the matrix g (or G)). The requirements (3.11) will further decrease the number of free parts of components as follows :

$$Aa) \qquad g_1^S + g_1^A = \tilde{g}_1^S + \tilde{g}_1^A \; ; \quad g_2^S + g_2^A = \tilde{g}_2^S + \tilde{g}_2^A ,$$

$$\text{i.e.} \quad g_1^S \neq 0, \; g_1^A = 0 \; ; \quad g_2^S \neq 0, \; g_2^A = 0 ,$$

$$q = q^S = q_1^S + i\, q_2^S \tag{3.16}$$

with the analogous result for the G matrix. Thus, the matrices q and G are symmetric, and the number of free parts of components is equal to $n(n+1)$.

Ab) $\overset{S}{q_1} + \overset{A}{q_1} = \overset{\sim S}{q_1} + \overset{\sim A}{q_1}$; $\overset{S}{q_2} + \overset{A}{q_2} = -(\overset{\sim S}{q_2} + \overset{\sim A}{q_2})$,

(3.17)

i. e. $\overset{S}{q_1} \neq 0,\ \overset{A}{q_1} = 0$; $\overset{S}{q_2} = 0,\ \overset{A}{q_2} \neq 0,\ q = \overset{S}{q_1} + i\overset{A}{q_2}$

with the analogous result for the G matrix. Thus, the matrix q_1 (G_1), being symmetric and the matrix q_2 (G_2) antisymmetric, the number of free parts of components is equal to n^2 .

Ba) $\overset{S}{q_1} + \overset{A}{q_1} = \overset{\sim S}{G_1} + \overset{\sim A}{G_1}$; $\overset{S}{q_2} + \overset{A}{q_2} = \overset{\sim S}{G_2} + \overset{\sim A}{G_2}$,

(3.18)

i. e. $\overset{S}{q_1} = \overset{S}{G_1},\ \overset{A}{q_1} = -\overset{A}{G_1}$; $\overset{S}{q_2} = \overset{S}{G_2},\ \overset{A}{q_2} = -\overset{A}{G_2}$,

and the matrices q and G are orthogonal matrices

$$\tilde{q} = q^{-1}\ , \quad \tilde{G} = G^{-1}\ ,$$

the number of free parts of components being equal to $n(n-1)$.

$$\text{Bb)}\quad \overset{S}{g_1} + \overset{A}{g_1} = \overset{\sim S}{G_1} + \overset{\sim A}{G_1} \;;\quad \overset{S}{g_2} + \overset{A}{g_2} = -(\overset{\sim S}{G_2} + \overset{\sim A}{G_2}), \tag{3.19}$$

$$\text{i.e.}\quad \overset{S}{g_1} = \overset{S}{G_1} \quad \overset{A}{g_1} = -\overset{A}{G_1} \;;\quad \overset{S}{g_2} = -\overset{S}{G_2},\; \overset{A}{g_2} = \overset{A}{G_2},$$

and the matrices g and G are unitary matrices

$$g^{+} = g^{-1}, \quad G^{+} = G^{-1},$$

the number of free parts of components being equal to n^2.

For the commutative vector components we see from (3.11) that the scalar product is commutative in cases $Aa)$ and $Ba)$, but in cases $Ab)$ and $Bb)$ it is Hermitian.

With the help of (3.11) we are able to express the scalar product of a vector multiplied by itself in terms of its components in the form

	a	b
A)	$a_i\, a^i$	$\overset{*}{a_i}\, a^i$
B)	$a^i\, a^i$	$a^{i*}\, a^i$

(3.20)

For a real vector space $a_i\,,\, a^i \in R$ (real quantities) this scalar product (3.20) is a real quantity; in case B it is always positive definite. For a complex vector space $a_i\,,\ a^i \in C$ (complex quantities) the scalar product is certainly positive definite in case $Bb)$ only.

In case $Bb)$ we define the norm of a vector as a positive real quantity

$$\|a\| = +\sqrt{a\,a} = +\sqrt{a^{i*}\,a^i}\,. \qquad (3.21)$$

The quantity (3.21) indeed possesses the three required properties :

1) $\|a\| \geq 0\,;\ = 0$ iff $a^i = 0,\ i = 1, 2, \ldots n$,

2) $\|(\lambda a)\| = +\sqrt{(\lambda a)^{<}{}_{>}(\lambda a)} = |\lambda|\ \|a\|$, because of (3.20).

3) Because of (3.21) and (3.20), for a real μ we have

$$0 \leq \|a + \mu b\|^2 = \|a\|^2 + 2\mu\, Re\,(a\,b) + \mu^2\ \|b\|^2. \quad (3.22)$$

Hence it follows that the discriminant D of the quadratic equation must satisfy the condition $D \leq 0$, which then leads to the Schwarz relation

$$Re\,(a\,b) \leq ||a|| \cdot ||b|| \tag{3.23}$$

Finally, taking $\mu = 1$ from the relations (3.22) and (3.23) we deduce the triangle inequality

$$||a + b|| \leq ||a|| + ||b|| \tag{3.24}$$

CHAPTER IV

The transformation of the basis vectors

Now, we make another choice of the basis vectors in the vector spaces $X^{<}$, $X_{<}$, ${}^{>}X$ and ${}_{>}X$ and use a prime to denote all quantities referring to the new systems of basis vectors, assuming all previously established connections to be valid for the primed systems too. In this way, with the help of the identity operators (2.1) in the primed systems, we obtain the connections of the basis

· vectors

	$e^i =$	$e_i =$	${}^i e =$	${}_i e =$
${}_{>}E'^{<}$	$(e^i \, {}_j e')\, e'^j$	$(e_i \, {}_j e')\, e'^j$	${}_j e' \,(e'^j \, {}^i e)$	${}_j e' \,(e'^j \, {}_i e)$
${}^{>}E'_{<}$	$(e^i \, {}^j e')\, e'_j$	$(e_i \, {}^j e')\, e'_j$	${}^j e' \,(e'_j \, {}^i e)$	${}^j e' \,(e'_j \, {}_i e)$

(4. 1)

Analogous relations are obtained by interchanging the role of primed and unprimed quanti - ties. The transformation coefficients, i. e. the components of the basis vectors in the other system, have to satisfy the relations obtained from (1. 6)

$$e^i \, {}_{>}E'^{<} \, {}_j e = \delta^i_j \quad , \quad e_i \, {}^{>}E'_{<} \, {}^j e = {}_i\delta^j \, , \tag{4.2}$$

$$e^i \, {}_{>}E'^{<} \, {}^j e = e^i \, {}^{>}E'_{<} \, {}^j e = g^{ij} , \; e_i \, {}_{>}E'^{<} \, {}_j e = e_i \, {}^{>}E'_{<} \, {}_j e = {}_i g_j \, ,$$

and a similar set of relations is valid for the inverse trans- formation, the determinants of the transformation coeffi- cients being assumed not to vanish.

The transformation law of the vector components is easily obtained with the help of (4. 1) from the representations of the same vector in both systems of basis vectors

$$({}_{>}a)' = {}_{>'}a = {}^{i}a'\,{}_{i}e' = {}_{>}E'^{<}\,{}_{>}a = {}_{i}e'\,e'^{i}\,{}^{k}a\,{}_{k}e = (e'^{i}\,{}_{k}e)\,{}^{k}a\,{}_{i}e' ,$$

$${}^{>'}a = {}^{>}E'_{<}\,{}^{>}a , \quad a_{<'} = a_{<}\,{}^{>}E'_{<} , \quad a^{<'} = a^{<}\,{}_{>}E'^{<} . \tag{4.3}$$

Analogous relations are obtained by interchanging the role of the primed and unprimed quantities.

The transformation properties of the valence raising and valence lowering operators (2. 4) follow immediately

$${}^{>}g'^{<} = {}^{>}E'_{<}\,{}^{>}g^{<}\,{}_{>}E'^{<} , \quad {}_{>}g'_{<} = {}_{>}E'^{<}\,{}_{>}g_{<}\,{}^{>}E'_{<} ,$$

$${}^{>}g^{<} = {}^{>}E_{<}\,{}^{>}g'^{<}\,{}_{>}E^{<} , \quad {}_{>}g_{<} = {}_{>}E^{<}\,{}_{>}g'_{<}\,{}^{>}E_{<} . \tag{4.4}$$

Hence we see that the relations (2. 7) are valid for the primed system, too.

The value of the scalar product (2. 11) is an invariant of the transformation (4. 1) (or (4. 3)), because we have

$$a'^{<}\,{}_{>}b' = a^{<}\,{}_{>}E'^{<}\,{}_{>}E'^{<}\,{}_{>}b = a^{<}\,{}_{>}b . \tag{4.5}$$

We could combine the transformations of the basis vectors i.e. after the transformation of the basis vectors e to the primed basis vectors e' we transform these primed basis vectors to new, double primed basis vectors e''. The latter transformation is again connected with an appropriate scheme (4.1) with ${}_{>}E''^{<} = {}_{i}e'' \; e''^{i}$ and ${}^{>}E''_{<} = {}^{i}e'' \; e''_{i}$ as identity operators. We have

$$e^{i} = e^{i} \, {}_{>}E'^{<}, \quad e'^{j} = e'^{j} \, {}_{>}E''^{<}, \quad e^{i} = e^{i} \, {}_{>}E'^{<} \, {}_{>}E''^{<} = e^{i} \, {}_{>}E''^{<}, \qquad (4.6)$$

and similar relations for other basis vectors. The relations (4.6) reflect the group character of the basis vector transformations, because for the transformation coefficients we explicitly have

$$(e^{i} \, {}_{j}e')(e'^{j} \, {}_{k}e'') = (e^{i} \, {}_{k}e'') \qquad (4.7)$$

and similar sets of relations for other basis vectors.

CHAPTER V

Tensors

The results obtained for vectors can be extended to more general objects, tensors, [1] in the following way. We take N vector spaces $X(\alpha)$ $(\alpha = 1, 2, \ldots, N)$, each of them belonging to one of the four kinds described previously. Each vector space $X(\alpha)$, the dimension dim $X(\alpha) = n_\alpha$, has its system of basis vectors $\mathbf{e}(\alpha)$ acting in the α space as indicated by (1.6). Then we construct the product vector space $X = X(1) \otimes X(2) \otimes \ldots \otimes X(N)$, the symbol $\otimes$ indicates that in any case the vector valences do not act one upon another. Thus, we assume that the factor vector spaces $X(\alpha)$ in the product space X are ordered in a sequence of β ket and γ bra spaces, $\beta + \gamma = N$. The role of basis vectors in the product space X is played by

$$\mathcal{N} = \prod_{\alpha=1}^{N} n_\alpha \qquad \text{quantities}$$

$$e = \prod_{\kappa=1}^{\beta} e(\kappa) \prod_{\lambda=1}^{\gamma} e(\lambda) ; \; e(\kappa) \in X_{ket}(\kappa) , \quad e(\lambda) \in X_{bra}(\lambda) . \tag{5. 1}$$

Each quantity represented by $\mathcal{N}$ components with the help of the $\mathcal{N}$ basis vectors (5. 1) in the product space X is a vector subjected to the axioms (1. 1) and multilinearly dependent on the basis vectors of the factor spaces $X(\alpha)$. We call such a quantity a tensor and specify its type by specifying the order and sequence of ket (and bra)-up and -down factor spaces in the product vector space

$$\text{tensor type} = \left(\begin{array}{c|c} \text{ket-up} & \text{bra-up} \\ \text{ket-down} & \text{bra-down} \end{array} \right) \equiv \left(\begin{array}{c|c} \beta_u & \gamma_u \\ \beta_d & \gamma_u \end{array} \right) \equiv (\beta, \gamma) \tag{5.2}$$

or explicitly indicating the sequences of the corresponding valences. The number $N = \beta + \gamma$ is called the rank of the tensor.

As an example we write the tensor

$$^{1>}_{2>}T^{<3}_{<4\,<5} = {}^{j}e(1)\; {}_{k}e(2) ({}^{k}_{j}T^{pr}_{\ell})\, e^{\ell}(3)\; e_p(4)\; e_r(5) \tag{5. 3}$$

where ${}^{j}e\,(1) \in {}^{>}X\,(1) \quad (j = 1, 2, \ldots, n_1)$ etc., the product space being

$$ {}^{1>}_{2>}X^{<3}_{\ <4\ <5} = {}^{>}X\,(1)\ {}_{>}X\,(2)\ X^{<}(3)\ X_{<}(4)\ X_{<}(5) \qquad (5.4a)$$

with the basis vectors

$$ {}^{j}_{k}e^{l}_{\ pr} = {}^{j}e\,(1)\ {}_{k}e\,(2)\ e^{l}(3)\ e_{p}\,(4)\ e_{r}\,(5) \qquad (5.4b)$$

linearly dependent on the basis vectors of the factor spaces. The type of the tensor (5.3) is indicated by

$$\left(\begin{matrix}1\\1\end{matrix}\middle|\begin{matrix}1\\2\end{matrix}\right) \equiv \left(\begin{matrix}1 & \cdot\\ \cdot & 1\end{matrix}\middle|\begin{matrix}1 & \cdot & \cdot\\ \cdot & 1 & 1\end{matrix}\right) \equiv (2|3) \qquad (5.4c)$$

and the number of its components is equal to $\mathcal{N} = n_1\, n_2\, n_3\, n_4\, n_5$.

Similarly to the vectors (Ch. I), the tensors in the same product space X can be added, subtracted and multiplied by the scalars ($\in K$).

Now, it is easy to define the direct tensor product as the tensor with components equal to the product of the corresponding factor tensor components, in the product space of the factor tensor spaces. Symbolically we write

$$T = T_1 \otimes T_2 \otimes \ldots \otimes T_m \,, \quad T_i \in X_i \,, \quad T \in X = \prod_{i=1}^{m} \otimes X_i \,, \tag{5.5a}$$

the type of the tensor T being

$$(\beta | \gamma) \equiv \left(\begin{array}{c|c} \beta_u & \gamma_u \\ \beta_d & \gamma_d \end{array} \right) \equiv \left(\begin{array}{c|c} \sum_i \beta_{ui} & \sum_i \gamma_{ui} \\ \sum_i \beta_{di} & \sum_i \gamma_{di} \end{array} \right) \tag{5.5b}$$

The rank of the tensor T is equal to the sum of factor tensor ranks.

As an example we write the direct tensor product explicitly

$$ {}_{1>}{}^{2>}A^{<3}{}_{<4} \otimes {}_{5>\,6>}B^{<7\,<8} = {}_{1>}{}^{2>}{}_{5>\,6>}C^{<3}{}_{<4}{}^{<7\,<8} = \tag{5.6a}$$

$$= {}_{i}e(1)\, {}^{j}e(2)\, {}_{m}e(5)\, {}_{n}e(6) \left({}_{j}^{i}A_{k}{}^{\ell mn} B_{pr} \right) e^{k}(3)\, e_{\ell}(4)\, e^{p}(7)\, e^{r}(8)$$

the type of the tensor (5.6a) being

$$(4|4) = \left(\begin{array}{c|c} 1 & 3 \\ 3 & 1 \end{array} \right) \equiv \left(\begin{array}{cccc|cccc} \cdot & 1 & \cdot & \cdot & 1 & \cdot & 1 & 1 \\ 1 & \cdot & 1 & 1 & \cdot & 1 & \cdot & \cdot \end{array} \right) = \left(\begin{array}{cc|cc} \cdot & 1 & 1 & \cdot \\ 1 & \cdot & \cdot & 1 \end{array} \right) + \left(\begin{array}{cc|cc} \cdot & \cdot & 1 & 1 \\ 1 & 1 & \cdot & \cdot \end{array} \right) \tag{5.6b}$$

Naturally, in the direct tensor product we could allow definite valences, when belonging to the same α bra and ket vector spaces, to act one upon another according to (1.6). In this way we obtain the contracted direct product, each contracted pair of valences (marked by a dash) decreasing the number of ket and bra valences by one. For instance, we contract the direct product (5.6a) in the following two manners:

$$ {}_{1>}^{2>}A^{\overline{<3}}{}_{<4} \otimes {}_{\overline{3>}\,6>}B^{<7\,<8} = {}_{i}e(1)\,{}^{j}e(2)\,{}_{n}e(6)\left({}^{i}{}_{j}A_{k}{}^{\ell\,kn}B_{pr}\right)e_{\ell}(4)\,e^{p}(7)\,e^{r}(8) \qquad (5.7a)$$

$$ {}_{1>}^{2>}A^{\overline{<3}}{}_{\overline{<4}} \otimes {}_{\overline{4>}\,\overline{3>}}B^{<7\,<8} = {}_{i}e(1)\,{}^{j}e(2)\left({}^{i}{}_{j}A_{k}{}^{\ell}\,{}_{\ell}g_{m}(4)\,{}^{mk}B_{pr}\right)e^{p}(7)\,e^{r}(8)$$

the types of tensors being

$$(3|3) = \left(\begin{array}{c|c} 1 & 2 \\ 2 & 1 \end{array}\right) = \left(\begin{array}{ccc|ccc} \cdot & 1 & \cdot & \cdot & 1 & 1 \\ 1 & \cdot & 1 & 1 & \cdot & \cdot \end{array}\right);$$

$$(2|2) = \left(\begin{array}{c|c} 1 & 2 \\ 1 & 0 \end{array}\right) = \left(\begin{array}{cc|cc} \cdot & 1 & 1 & 1 \\ 1 & \cdot & \cdot & \cdot \end{array}\right), \qquad (5.7b)$$

respectively.

Applying the results of Chapter II to the factor spaces $X(\alpha)$ we determine the one-to-one correspondence between the tensors differing only in the same bra (ket)-up and down valences. We illustrate this fact by the example (5.3)

$$
\begin{aligned}
{}^{1>\,2>}T^{<3\,<4\,<5} &= {}^{1>}\left({}_{2>}\right) T^{<3} \left({}_{<4}\right)\left({}_{<5}\right) = \\
&= {}^{i}e(1)\,{}^{k}e(2)\;{}_{k}q_{\bar{k}}(2)\;{}_{i}{}^{\bar{k}}T_{\ell}^{\bar{p}\bar{r}}\;{}_{\bar{p}}q_{p}(4)\,{}_{\bar{r}}q_{r}(5)\,e^{\ell}(3)\,e^{p}(4)\,e^{r}(5)\,.
\end{aligned}
\tag{5.8}
$$

Here we used the abbreviations

$$
\left({}_{\alpha>}\right) = \left({}^{>}q^{<}(\alpha)_{\alpha>}\right) \qquad \left({}_{<\alpha}\right) = \left({}_{<\alpha}{}^{>}q^{<}(\alpha)\right) \tag{5.9}
$$

to indicate the raising of the α -valence. In an analogous way the symbols $\left({}^{\alpha>}\right)$, $\left({}^{<\alpha}\right)$ mean the lowering of the α valence. The relation between the tensor components (5.8) follows immediately.

The connection between the bra and ket vector forms (Chapter III) can be extended to tensors, and a one-to-one correspondence can be established between the tensors in the product vector space and in its mirror-image vector space (case A) or mirror-image vector space with changed character (up and down) of the valences of every factor space (case B). From the requirement that the associated quantities of the associated quantities should be identical with the original ones we can draw the conclusion that in such a correspondence there are

A) tensors whose bra (ket) up and down valences are symmetrically changed into ket (bra) up and down valences with a) components unchanged, b) components conjugate complex,

B) tensors whose bra (ket) up and down valences are changed into ket (bra) down and up valences with a) components unchanged, b) components conjugate complex.

We illustrate these statements by the example (5. 3), the associated tensors being

$$\begin{aligned}
&Aa)\quad {}_{5>\,4>}{}^{3>}B_{<2}{}^{<1} \;;\quad {}^{rp}{}_{\ell}B^{k}{}_{j} \to {}_{j}{}^{k}T_{\ell}{}^{pr} ,\\
&Ab)\quad {}_{5>\,4>}{}^{3>}C_{<2}{}^{<1} \;;\quad {}^{rp}{}_{\ell}C^{k}{}_{j} \to {}_{j}{}^{k}T_{\ell}{}^{pr\,*}\\
&Ba)\quad {}^{5>\,4>}{}_{3>}D^{<2}{}_{<1} \;;\quad {}_{rp}{}^{\ell}D_{k}{}^{j} \to {}_{j}{}^{k}T_{\ell}{}^{pr} ,\\
&Bb)\quad {}^{5>\,4>}{}_{3>}F^{<2}{}_{<1} \;;\quad {}_{rp}{}^{\ell}F_{k}{}^{j} \to {}_{j}{}^{k}T_{\ell}{}^{pr\,*} .
\end{aligned} \tag{5.10}$$

It is easily seen that in the associated direct tensor product the reversed order of the associated factor tensors has to be taken.

When we change the basis vectors in each of the factor spaces, we obtain the transformation law for tensors in the product space by applying the identity operators in each factor space separately. Here we use the abbreviation

$$ {}_{\alpha>'} \equiv \left({}_{>}E'(\alpha)^{<}{}_{\alpha>} \right) \qquad {}^{\alpha>'} \equiv \left({}^{>}E'(\alpha)_{<}{}^{\alpha>} \right) \tag{5.11} $$

and analogous expressions for other valences. The connection of the tensor representation in the primed (original) system of basis vectors with that in the original (primed) system is easily given in a compact form, as illustrated by the example (5.3)

$$ \left({}^{1>}{}_{2>} T^{<3}{}_{<4\,<5} \right)' = {}^{1>'}{}_{2>'} T^{<3'}{}_{<4'\,<5'} \quad , \tag{5.12a} $$

or explicitly written (for brevity we omit α)

$$ {}^{\bar{j}}e'\,{}_{\bar{k}}e' \left({}_{\bar{j}}{}^{\bar{k}} T'_{\bar{\ell}}{}^{\bar{p}\bar{r}} \right) e'^{\bar{\ell}}\, e'_{\bar{p}}\, e'_{\bar{r}} = \tag{5.12b} $$

$$ = {}^{\bar{j}}e'\,{}_{\bar{k}}e' \left[(e'_{\bar{j}}{}^{j}e)(e'^{\bar{k}}{}_{k}e)\,{}_{j}{}^{k}T_{\ell}{}^{pr}\,(e^{\ell}{}_{\bar{\ell}}e')(e_{p}{}^{\bar{p}}e')(e_{r}{}^{\bar{r}}e') \right] e'^{\bar{\ell}}\, e'_{\bar{p}}\, e'_{\bar{r}} \, . $$

Hence the transformation law

for the tensor components follows at once.

Of special importance is a particular case when all the factor spaces $X(\alpha)$ in the product space belong to the same α, i.e. they are ${}_{>}X$, ${}^{>}X$, $X_{<}$, $X^{<}$ of the same dimension n and their valences mutually interact. Then, we could omit α and abbreviate the notation as illustrated by the example of the tensor (compare with (5.3))

$$ {}^{>}_{>}T^{<}{}_{<\,<} = {}^{j}e\; {}_{k}e \left({}_{j}{}^{k}T_{\ell}{}^{pr} \right) e^{\ell}\, e_{p}\, e_{r} \qquad (5.13) $$

in the product space $X = {}^{>}X \otimes {}_{>}X \otimes X^{<} \otimes X_{<} \otimes X_{<}$.

All the results obtained above can be easily specialized to this particular case.

CHAPTER VI

Remarks

We add a few remarks to the described approach to the vector and tensor algebra.

a) The notation introduced in this paper could be regarded as a contribution to the "direct" system of notation in the sense of Schouten [3] . Its usefulness for the vector and tensor analysis will be shown in a separate paper.

b) The results of Chapter III and the related matter in Chapter V dealing with the connections of the bra and ket forms of vectors and tensors may be regarded as a contribution to the theory of "hybrid quantities" in the sense of Schouten [3] .

c) We see from Chapter I that the assumptions (1.6) play the central role. The whole scheme depends on the valence raising and lowering operators, or as we could call them (Chapter V) on fundamental tensors, and a particular case is characterized by their explicit expressions. For real g (and G) we could distinguish the so-called Euclidean and Minkowskian case, the scalar product (3.20) being definite in the first case and indefinite in the second. Also, we could call the system of basis vectors to be Cartesian when g (and G) are diagonal matrices with diagonal components equal to ± 1 .

We add a few particular cases :

1) We assume the fundamental tensors to be of the form (2.8) where the difference between cases A and B disappears Because of (1.6) the basis vectors are Cartesian orthonormal unit vectors (their norm (3.21) equal to + 1). For a real vector space we have the Euclidean case, because the difference

between cases a) and b) disappears and the norm (3.21) of each vector is positive. For a complex vector space the difference between cases a) and b) exists and the Euclidean case is realized in case b) only.

2) For the Cartesian basis vectors, with all diagonal components of the fundamental tensor equal to - 1 (+ 1) except one component equal to + 1 (- 1), the real vector space in case A is Minkowskian, due to the indefinite scalar product (3. 20) $A a$), b). It is interesting to note that case 1) a) for $n = 4$, with three real (imaginary) vector components and the fourth imaginary (real) component will be identical to the mentioned Minkowskian case. This fact was reflected in the formulations of the special theory of relativity.

3) The two-dimensional complex vector space with the fundamental tensors

$$g \equiv \begin{pmatrix} 0 & 1 \\ -1 & 0 \end{pmatrix}, \quad G \equiv \begin{pmatrix} 0 & -1 \\ 1 & 0 \end{pmatrix}, \tag{6.1}$$

represents the basis for the 2-spinor theory. The four-dimensional complex vector space which is the direct sum of two 2-spinor spaces is the basis for the 4-spinor theory.

4) The spherical basis vectors are defined with the

help of ordinary three-dimensional Cartesian orthonormal unit vectors i , j *and* k in the following way

$$e_{+1} = \frac{<i - i<j}{\sqrt{2}} , \; e_0 = <k , \; e_{-1} = -\frac{<i + i<j}{\sqrt{2}} ,$$

$$_{+1}e = \frac{i> + i\, j>}{\sqrt{2}} , \; _0e = k> , \; _{-1}e = -\frac{i> - i\, j>}{\sqrt{2}} . \qquad (6.2a)$$

The spherical basis vectors are therefore orthonormal unit vectors too, the fundamental tensor being of the form (2. 8)

$$e_{\nu\,\mu}e = {}_\nu g_\mu = {}^\nu g^\mu = {}_\nu\delta_\mu , \quad \nu, \mu = -1, 0, +1 . \qquad (6.2b)$$

Thus, there is no difference between cases A and B , and from the definition of spherical basis vectors we conclude that case *b)* (with a change in the sign of components) is realized. In fact, the transition from the Cartesian basis vectors to the spherical ones represents the transformation of basis vectors (Chapter IV) in a Euclidean space.

d) In spite of the fact that we developed the described scheme for finite-dimensional vector spaces $X(n)$ it can usefully be applied in a discussion of more general cases, i. e. in infinite-dimensional spaces $X(\infty)$ and $X(C)$. Here we do not give a full and rigorous report but indicate some plausible changes to be made with some illustrations for application.

First, we have to interpret the summation over double indices as sums with an infinite number of terms for $n = \infty$ and as integrals over double indices for the continuum. Thus, for instance we write

$$n \text{ finite} \qquad X_<(n)\,,\ a_< = a^i e_i = \sum_{i=1}^{n} a^i e_i\,,$$

$$n \text{ infinite} \qquad X_<(\infty)\,,\ a_< = a^i e_i = \sum_{i=1}^{\infty} a^i e_i\,, \tag{6.3}$$

$$\text{continuum} \qquad X_<(C)\,,\ a_< = a^x e_x = \int_R a^x e_x\, dx = \int_R a(x)\, e_x\, dx\,.$$

Here, we wrote the "continuous index" with "x" ; $a^x \equiv a(x)$ is the component dependent on the continuous index, i. e. a function of the continuous index as variable, R means the domain of the variable x , the vector $a_<$ symbolizes, in fact, the totality of function values $a(x)$. The other three vector forms (1. 2) can be interpreted analogously.

The basis vector scalar products for $n = \infty$ are of the form (1. 6). Here the indices i and j take all integer (natural integer) values and the matrices g and G (2. 7) are of infinite order.

For the continuum the scheme (1. 6) has to be generalized to this form

$$e^{x}{}_{y}e = \delta(x-y)\,, \qquad e_{x}{}^{y}e = \delta(x-y)\,,$$

$$e^{x}{}^{y}e = {}^{x}q^{y}\ (= G(x,y))\,, \qquad e_{x}{}_{y}e = {}_{x}q_{y}\ (= q(x,y))\,, \tag{6.4}$$

where the "Dirac function" and functions of two variables $G(x,y)$ and $q(x,y)$ are used to substitute the Kronecker symbol δ and matrix components $\overset{ij}{q}$ and $_{i}q_{j}$ in (1.6).

To illustrate the application, we add two simple examples.

1) We assume that the particular case (2.8) for both spaces $X(\infty)$ and $X(C)$ is realized, i.e.

$$_{i}q_{j} = \overset{ij}{q} = \underset{ij}{\delta}\,, \qquad {}_{x}q_{y} = \overset{xy}{q} = \delta(x-y)\,. \tag{6.5}$$

Thus, for the two spaces the difference between cases A and B disappears.

We introduce a one-to-one correspondence between the vectors of the two spaces $X(\infty)$ and $X(C)$ by means of the linear operators

$$_{i>}T_{<x} = {}_{i}e\ {}_{i}T_{x}\ e_{x}\,, \qquad {}_{x>}\overline{T}_{<i} = {}_{x}e\ {}_{x}\overline{T}_{i}\ e_{i}\,. \tag{6.6}$$

The associated vectors being written with a dash, we have for the basis vectors

$$\bar e_i = e_i\, {}_{i>}T_{<x} = {}_iT_x\, e_x \quad , \quad {}_i\bar e = {}_{x>}\bar T_{<i}\, {}_ie = {}_x\bar T_i\, {}_xe \, ,$$

$$\bar e_x = e_x\, {}_{x>}\bar T_{<i} = {}_x\bar T_i\, e_i \quad , \quad {}_x\bar e = {}_{i>}T_{<x}\, {}_xe = {}_iT_x\, {}_ie \, . \tag{6.7}$$

The requirement that the associated vector of the associated vector should be the original one (e. g. $\bar{\bar e}_i = e_i$, $\bar{\bar e}_x = e_x$) leads to the conclusion that the operators (6. 6) have to be inverse operators one to the other. Thus, they satisfy the relations

$${}_{i>}T_{<x}\, {}_{y>}\bar T_{<j} = {}_ie\, {}_iT_x\, {}_x\bar T_j\, e_j = {}_ie\, {}_i\delta_j\, e_j = {}_{i>}E_{<i} \, ,$$

$${}_{x>}\bar T_{<i}\, {}_{j>}T_{<y} = {}_xe\, {}_x\bar T_i\, {}_iT_y\, e_y = {}_xe\, \delta(x-y)\, e_y = {}_{x>}E_{<x} \tag{6.8a}$$

As a consequence of the relations (6. 7) and (6. 8a) we easily find the correspondence between the identity operators

$${}_{i>}\bar E_{<i} = {}_i\bar e\, \bar e_i = {}_{x>}E_{<x} \quad , \quad {}_{x>}\bar E_{<x} = {}_x\bar e\, \bar e_x = {}_{i>}E_{<i} \, . \tag{6.9}$$

We now determine the corresponding vectors with the help of the operators (6. 6)

$$a_{<x}\, {}_{x>}\bar T_{<i} = a_x\, {}_x\bar T_i\, e_i = a_i\, e_i = a_{<i} \, , \tag{6.10a}$$

$$a_{<i}\ {}_{i>}T_{<x} = a_i\ {}_iT_x\ e_x = a_x\ e_x = a_{<x} \quad ,$$

and the analogous relations for the ket vector forms. We point out that the scalar product is invariant

$$a_{<i}\ {}_{i>}b = a_{<i}\ {}_{i>}T_{<x}\ {}_{x>}\overline{T}_{<i}\ {}_{i>}b = a_{<x}\ {}_{x>}b \ . \qquad (6.10b)$$

If case $b)$ is taken into consideration, from the relations (6. 7) for the operator components (6. 6) it follows

$${}_iT_x = {}_x\overline{T}_i^{\,*} = (e_i\ {}_x\overline{e}) = (\overline{e}_x\ {}_ie)^* = \Phi_i(x) \ . \qquad (6.11)$$

The functions $\Phi_i(x)$ represent an infinite orthonormal complete set of functions as shown from (6. 8a) when written in the conventional manner (6. 3)

$$\begin{aligned} {}_iT_x\ {}_x\overline{T}_j &= \int_R \Phi_j^*(x)\ \Phi_i(x)\ dx = {}_i\delta_j \ , \\ {}_x\overline{T}_i\ {}_iT_y &= \sum_i \Phi_i^*(x)\ \Phi_i(y) = \delta(x-y) \ , \end{aligned} \qquad (6.8b)$$

where R is the interval of the variable x .

In case *b*) the relations (6.10a) take these forms in the usual notation $(a_x \equiv a(x))$

$$a_{<i} = \int_R a(x)\, \Phi_i^*(x)\, dx\, e_i = a_i e_i\,, \quad a_i = \int_R a(x)\, \Phi_i^*(x)\, dx\,,$$

(6.12)

$$a_{<x} = \int_R a_i\, \Phi_i\,(x)\, e_x\, dx = \int_R a(x)\, e_x\, dx\,, \quad a(x) = a_i\, \Phi_i\,(x)\,.$$

Thus, the relation (6.10a) or (6.12) represents the series expansion of the function $a(x)$ by means of the orthonormal complete set of functions (6.11), the totality of the function values being symbolized by $a_<$

The natural requirement of the finite vector norm (3.20) is written as

$$\|\mathbf{a}\|^2 = a_< {}_>a = a_x\, e_x\, {}_y a\, {}_y e = \int_R |a(x)|^2 dx < \infty$$

(6.13)

$$= a_i\, e_i\, {}_j a\, {}_j e = \sum_i |a_i|^2 < \infty\,.$$

It is equivalent to the requirement that the function $a(x)$ is a quadratically integrable function in $X(C)$ and a quadratically summable function in $X(\infty)$.

The original Fourier series with the orthonormal complete set of functions

$$\Phi_k(x) = e^{ikx}/\sqrt{2\pi}\,, \quad k = \text{ integer } , \quad R \equiv [\alpha, 2\pi + \alpha] \tag{6.14}$$

represents a realization of the above mentioned case.

2) If we had taken $X(C)$ instead of $X(\infty)$, in (6.5), we would have developed quite an analogous discussion to the preceding one. In this case the continuous indices ξ and η would have substituted the discrete indices i and j and the summation over discrete indices would have been substituted by integration over the corresponding continuous indices. In fact, here we have a transformation (Chapter IV) of the basis vectors $e_x \in X(C)$ to the new basis vectors $e_\xi \in X(C)$, the operator components (6.11) ${}_\xi T_x = {}_x \overline{T}^*_\xi = \Phi(\xi, x)$ being the transformation coefficients, The relations analogous to (6.11), (6.12) and (6.13) have the forms

$$a_< = a_\xi \, e_\xi = (a_< \, {}_\xi e)\, e_\xi = a_x \, e_x = (a_< \, {}_x e)\, e_x$$

$$= a_x \,(e_x \, {}_\xi e)\, e_\xi = a_\xi (e_\xi \, {}_x e)\, e_x \;, \tag{6.15}$$

$$a_< = \int_{R_\xi}\int_{R_x} a(x)\,\Phi^*(\xi, x)\, dx\, d\xi \; e_\xi \,, \quad A(\xi) = \int_{R_x} a(x)\,\Phi^*(\xi, x)\, dx = (a_\xi),$$

$$a_< = \int_{R_x}\int_{R_\xi} A(\xi)\,\Phi(\xi,x)\,d\xi\,dx\,e_x\,, \quad a(x) = \int_{R_\xi} A(\xi)\,\Phi(\xi,x)\,d\xi = (a_x)\,,$$

$$\|a\|^2 = a_< {}_> a = \int_{R_x} |a(x)|^2\,dx = \int_{R_\xi} |A(\xi)|^2\,d\xi < \infty\,,$$

while the transformation coefficients satisfy (6. 8b)

$$\int_{R_x} \Phi^*(\eta,x)\,\Phi(\xi,x)\,dx = \delta(\eta-\xi)\,, \int_{R_\xi} \Phi^*(\xi,x)\,\Phi(\xi,y)\,d\xi = \delta(x-y)\,. \tag{6.16}$$

Taking the transformation function to be

$$\Phi(\xi,x) = e^{i\xi x}/\sqrt{2\pi}\,, \quad R_x \equiv (-\infty,+\infty),\quad R_\xi \equiv (-\infty,+\infty) \tag{6.17}$$

the foregoing formulas represent the Fourier integral and transform for the functions $a(x)$ $(A(\xi))$.

It is easy to extend this reasoning to the case of several variables, i. e. to series expansions, integral representations and integral transforms for functions of several variables starting from the scheme for tensors (Chapter V).

Evidently, the introduced

one-to-one correspondence (6.6) between the spaces $X(\infty)$ and $X(C)$ would find its deeper sense in the light of the well-known theorem [4] : Any complex (real) separable Hilbert space H is isometric and isomorphic to the complex (real) space ℓ_2 and, consequently, all complex (real) separable H spaces are isometric and isomorphic one to another.

We are not going here to discuss many possible applications of the described approach for the infinite dimensional case, but we only mention that this approach could be useful in many fields, e.g. theory of integral equations, quantum theory etc.

e) We could go over from the introduced notation with four kinds of indices to the notation with unilateral indices in several consistent manners. First, we have to take into account two possibilities of collecting unilaterally the indices on the right- and left-hand side. Further, in each of these cases the order of unilateral indices can be taken either bra, ket or ket, bra. In addition, the original tensor component and the corresponding component with unilateral indices must transform in the same way so that the correspondence established in one system of basis vectors remains conserved in any other system. Therefore, since the transformation coefficients are scalar products of the respective basis vectors, we have to take into account the results (3.11) and distinguish cases *a)* and *b)*. Finally, the relations (5.10), which connect the components of a tensor and of its associated tensor, must be conserved.

We illustrate different possibilities for case *a)* on the example (5.13) and write down the components with unilateral indices corresponding to the tensor component ${}_{j}^{k}T_{l}^{pr}$

a)	ket , bra right	ket , bra left	bra , ket right	bra , ket left
A)	$T_{j}{}^{k}{}_{l}{}^{pr}$	${}_{j}{}^{k}{}_{l}{}^{pr}T$	$T_{l}{}^{pr}{}_{j}{}^{k}$	${}_{l}{}^{pr}{}_{j}{}^{k}T$
B)	$T^{j}{}_{kl}{}^{pr}$	${}_{j}{}^{kl}{}_{pr}T$	$T_{l}{}^{prj}{}_{k}$	${}^{l}{}_{prj}{}^{k}T$

(6.18a)

In case *b)* the correspondence could be established only for tensors which are the direct product of vectors

$$ {}^{>}_{>}T^{<}_{<<} = {}^{>}a \, {}_{>}b \, c^{<} d_{<} f_{<} = {}^{j}e \, {}_{k}e \, {}_{j}a \, {}^{k}b \, c_{l} \, d^{p} f^{r} e^{l} e_{p} e_{r} \tag{6.19} $$

With the abbreviation

$$ {}_{\bar{j}}a = {}_{j}a^{*} $$

for the conjugate complex value we can compose an analogous scheme for tensors of the structure (6. 19) by recalling the rules for the change of vector indices in case (3. 9b)

b)	ket , bra		bra , ket	
	right	left	right	left
A)	$T_{\bar{j}}{}^{\bar{k}}{}_{\ell}{}^{pr}$	${}_{j}{}^{k}{}_{\bar{\ell}}{}^{\bar{p}\bar{r}}T$	$T_{\ell}{}^{pr}{}_{\bar{j}}{}^{\bar{k}}$	${}_{\bar{\ell}}{}^{\bar{p}\bar{r}}{}_{j}{}^{k}T$
B)	$T^{\bar{j}}{}_{\bar{k}\ell}{}^{pr}$	${}_{j}{}^{k\bar{\ell}}{}_{\bar{p}\bar{r}}T$	$T_{\ell}{}^{prj}{}_{\bar{k}}$	${}^{\bar{\ell}}{}_{\bar{p}\bar{r}j}{}^{k}T$

(6. 18b)

References

[1] H. J. Kowalsky: Lineare Algebra, Berlin 1963

W. H. Greub : Multilinear Algebra, Berlin 1967

[2] P. A. M. Dirac : The principles of quantum mechanics, 3. ed., Oxford 1949

[3] J. A. Schouten : Ricci-Calculus, 2. ed., Berlin 1954

[4] L. A. Ljusternik-V. I. Sobolev : Elementi funkcionalnovo analiza, Moskva 1965

N. S. Ahiezer-I. M. Glazman : Teoria linejnih operatorov v Hilbertovom prostranstve, Moskva 1966

K. Yosida : Functional Analysis, 2 ed., Berlin 1966

A CONTRIBUTION TO THE VECTOR AND TENSOR ANALYSIS I

In a preceding paper [1] we developed a simple approach to the vector and tensor algebra, whose basic characteristic was the equivalence of covariant, contravariant, bra and ket forms. Based on this feature, the vector and tensor analysis for vector and tensor fields is built up in the present paper, the n-dimensional vector spaces being connected with the points of an m-dimensional parameter manifold.

In Chapter I a one-to-one correspondence of the quantities $A(P) \in X(P)$, $A(P)_a \in X(Q)$, the so-called parallel displacement, is introduced for any pair of neighbouring points $P(x^k)$, $Q(x^k + dx^k)$ of the manifold. The definitions of the absolute, covariant and parallel displacement differentials (derivatives) follow, the respective sum and product rules having the same structure as those for ordinary differentials (partial derivatives). The absolute, covariant and parallel displacement differentials (derivatives) of every field quantity are easily determined after the absolute differentials of the basis vectors and of the scalar functions of coordinates have been given explicitly. The properties of the coefficients of connection, introduced in

the absolute differentials of the basis vectors, are extensively investigated in Chapter II, especially their connections with the fundamental tensor for cases $A\,a), b), B\,a), b)$ [1]
In Chapter III the second-order difference between the parallelly displaced quantities $A(P)_{Q_2 Q} - A(P)_{Q_1 Q}$, which is equal to the commutator of the absolute differentials $[\Delta]\, A(Q) = (\Delta_1 \Delta_2 - \Delta_2 \Delta_1)\, A(Q)$ is examined. The sum and product rules for the commutators of the absolute, covariant and parallel displacement differentials (derivatives) are found to have the same structure as those for the respective differentials (partial derivatives). The commutators of the absolute differentials (derivatives) of the vector and tensor fields are easily expressed in terms of the known basis vector commutators.

Introduction

In a preceding paper [1] we considered the vector space $X(P)$ connected with a specified set of parameter values, i.e. with a determined point $P(x^k)$

of an m -dimensional parameter manifold M . Now we assume the parameters to be real and to vary continuously in a certain domain Ω . We call a coordinate line in the domain Ω a set of points with only one definite parameter - coordinate - continuously varying, the other parameters remaining unchanged. Then every point $P(x^k) \in \Omega$ is the intersection of m coordinate lines.

The points $P \in \Omega$ can be labelled in an equivalent manner in another system of coordinate lines,. i.e. with another choice of the parameters x'^i We assume that there is one-to-one correspondence between all possible parameter choices so that for the coordinates the following relations exist

$$x'^i = x'^i(x^k) \, , \quad x^k = x^k(x'^i) \quad , \; x'^i, x^k \in C^N \, ,$$

$$(i, k = 1, 2, \ldots m) \tag{1}$$

$$d x'^i = \frac{\partial x'^i}{\partial x^k} d x^k \quad , \quad d x^k = \frac{\partial x^k}{\partial x'^i} d x'^i \quad .$$

The class of functions C^N is composed of all continuous functions with continuous partial derivatives up to the N -th order included, N being determined adequately.

Now, over the domain Ω we

define the vector (tensor) fields as sets of an adequate number of scalar functions $f_\alpha(x^k) \in C^N$ - the components of the vectors (tensors). The respective functional dependence on the coordinates x^k in a chosen system of coordinate lines for a given α is the same for each point $P(x^k) \in \Omega$. The system of basis vectors $\mathbf{e}(P) \in X(P)$, for every $P \in \Omega$ where the vector (tensor) fields are represented in the described way, are termed systems of the corresponding basis vectors of the vector spaces $X(P)$, $P \in \Omega$. Thus, a vector (tensor) field can be symbolically written as for example

$$
\begin{gathered}
{}_{>}a(P) = {}^{i}a(P)\, {}_{i}e(P) = {}^{i}a(x^k)\, {}_{i}e(P)\,, \\
{}^{>}q^{<}(P) = {}^{i}e(P)\, {}_{i}q_j(P)\, e^j(P) = {}^{i}e(P)\, {}_{i}q_j(x^k)\, e^j(P)\,,
\end{gathered} \tag{2}
$$

where ${}^{i}a(P)$ $({}_{i}q_j(P)) \in C^N$ represents, in a system of coordinate lines, the same function of coordinates for all $P \in \Omega$ (the coordinate lines (1) can be taken arbitrarily); the systems of the corresponding basis vectors being ${}_{i}e(P)$, ${}^{i}e(P)$, $e^i(P)$ and $e_i(P)$, $P \in \Omega$, respectively.

When we wish to represent the vector (tensor) fields in other systems of the corresponding basis vectors, we have to recall that the vector (tensor) character of the respective set of their vector (tensor) components in $X(P)$ is determined with the help of transformation laws for the basis vectors ([1], Chapter IV). In order to conserve the field characteristics it is necessary that the respective transformation coefficients are the same functions of coordinates for all $P \in \Omega$, which moreover satisfy the relations ([1], (4.2)) and manifest the group property ([1], (4.7)).

We can separately apply the results of the preceding paper [1] to each of the vector spaces $X(P)$, $P(x^k) \in \Omega$. The question arises how to connect these results obtained separately, in the case of vector and tensor fields when the corresponding n-dimensional vector spaces $X(P)$ and $X(Q)$ refer to different points $P, Q \in \Omega$ of the m-dimensional parameter manifold M.

CHAPTER I

Absolute, covariant and parallel displacement differential (derivative)

Absolute differential (derivative)

We start from a specified n-dimensional vector space $X(P)$ (for instance ${}^{>}X(P)$) with a system of basis vectors $\mathbf{e}(P)$ and from the vector space $X(Q)$ (${}^{>}X(Q)$) with a system of the corresponding basis vectors $\mathbf{e}(Q)$, the points $P, Q \in \Omega$ being labelled in the same system of coordinate lines and differing infinitesimally in the coordinates: $P(x^k)$ and $Q(x^k + dx^k)$. To connect the vector spaces $X(P)$ and $X(Q)$ means to connect all quantities referring to them. We therefore introduce the linear operators :

$$ {}^{P>}O_{<Q} = {}^{i}e(P)\, {}_{i}O^{j}\, e_{j}(Q), \quad {}_{P>}O^{<Q} = {}_{i}e(P)\, {}^{i}O_{j}\, e^{j}(Q), $$

(1. 1a)

$$ {}^{Q>}\bar{O}_{<P} = {}^{i}e(Q)\, {}_{i}\bar{O}^{j}\, e_{j}(P), \quad {}_{Q>}\bar{O}^{<P} = {}_{i}e(Q)\, {}^{i}\bar{O}_{j}\, e^{j}(P), $$

which should realize the connection of every vector $\mathbf{a}(P)$ ($\mathbf{a}(Q)$) in the vector space $X(P)$ ($X(Q)$) with a determined vector $a(P)_Q$ $(a(Q)_P)$ in the vector space $X(Q)$ $(X(P))$. Thus, for instance, we have

$$a_<(P)_Q = a_<(P)\,{}^{P>}O_{<Q} = a^i(P)\,{}_iO^i\,e_i(Q)\,,$$

$$a_<(Q)_P = a_<(Q)\,{}^{Q>}\bar{O}_{<P} = a^i(Q)\,{}_i\bar{O}^i\,e_i(P)\,,$$

and three similar relations for three other forms of vectors.

We require that the operators (1. 1a) should have the following properties :

a) The operators have to become the corresponding identity operators for $P \equiv Q$,

b) the components of the operators should be of the first order in the coordinate differences, i.e. they should contain only the first order terms in the parameter differentials (the points $P(x^k)$, $Q(x^k + dx^k)$ being the neighbouring points),

c) if only the first order terms in the parameter differentials are taken into account, the operators O and $\bar{O}$ should be inverse operators one to the other..

The one-to-one correspondence of the vectors produced by the described linear operators (1. 1a) with the properties a), b), c) we call the parallel displacement from one point to the other point, the operators themselves are called the parallel displacement operators and the vector $\mathbf{a}(P)_Q$ $(\mathbf{a}(Q)_P)$ the parallelly displaced vector $\mathbf{a}(P)$ $(\mathbf{a}(Q))$ from the point P (Q) to the point Q (P).

To satisfy the requirements a) and b), the operator components must have these explicit forms

$$
\begin{aligned}
{}_iO^j &= {}_i\delta^j - (\Gamma^j)_{ki}\, dx^k, \quad {}^iO_j = {}^i\delta_j - (\Gamma_j)^i_k\, dx^k, \\
{}_i\bar{O}^j &= {}_i\delta^j - {}^j({}_i\Gamma)_k\, dx^k, \quad {}^i\bar{O}_j = {}^i\delta_j - {}_j({}^i\Gamma)_k\, dx^k . \qquad (1.1b)
\end{aligned}
$$

The four sets of mn^2 (the index k running from 1 to m the indices i and j from 1 to n), i. e. $4mn^2$ quantities Γ are called the coefficients of connection, because they realize a definite connection of the vector spaces $X(P)$ and $X(Q)$ if given explicitly. They are functions of the point coordinates and consequently depend on the choice of the coordinate lines (1) as well as on the systems of the corresponding basis vectors. With $\Gamma \in C^N$, $N \geq 1$ in accordance with $b)$ the coefficients of connection can be taken in (1.1b) as functions of the coordinates x^k of the point P, i. e. $\Gamma = \Gamma(P) = \Gamma(x^k)$. The properties of the coefficients of connection will be discussed in detail in Chapter II.

The requirement c) is expressed in the form

$$ {}^{P>}O_{<Q} \; {}^{Q>}\bar{O}_{<P} = {}^{P>}E_{<P} + {}^{P>}O(2)_{<P} \, , $$

$$ {}_{P>}O^{<Q} \; {}_{Q>}\bar{O}^{<P} = {}_{P>}E^{<P} + {}_{P>}O(2)^{<P} \, , $$

and two analogous relations where O and $\bar{O}$ are exchanged. Here $O(2)$ mean operators with components of at least the second order terms in the coordinate differentials. With the help of (1. 1b) from these relations we obtain the following important conclusions

$$ (\Gamma_{i})^{i}_{k} + {}_{i}({}^{i}\Gamma)_{k} = 0 \, , $$

$$ (\Gamma^{i})_{ki} + {}^{i}({}_{i}\Gamma)_{k} = 0 \, , \qquad (1.\,2a) $$

In fact, the relations (1. 2a) reduce the number of coefficients of connection in (1. 1b) from $4\,m\,n^2$ to $2\,mn^2$.

Further, by an easy calculation from (1. 1), (1. 2a) and ([1] , (2. 7)) we find the relation

$$ {}^{P>}O_{<Q} \, {}_{Q>}\bar{O}^{<P} = {}^{P>}O_{<Q} \, {}_{Q>}q_{<Q} \, {}^{Q>}\bar{O}_{<P} \, {}^{P>}q^{<P} = {}^{P>}q(P)^{<P} + {}^{P>}O(2)^{<P} . $$

With the help of (1. 1) the left-hand side of this equation directly gives the expression

$$ {}^{P>}O_{<Q} \, {}_{Q>}\bar{O}^{<P} = {}^{P>}q^{<P} + {}^{i}e(P)\left[\partial_k \, {}_{i}q_{i} - (\Gamma^{r})_{ki} \, {}_{r}q_{i} - {}_{i}q_{r} \, {}_{i}({}^{r}\Gamma)_{k}\right]_P dx^k e^i(P) + $$
$$ + \, {}^{P>}O(2)^{<P} $$

where we assumed that ${}_{i}q_{j}\ (x^{k})\ \in C^{N}, N \geq 1$. By comparing the two obtained expressions we conclude that the factors of the coordinate differentials in the second relation have to vanish

$$\partial_{k}\ {}_{i}q_{j} = ({\Gamma^{r}})_{ki}\ {}_{r}q_{j} + {}_{i}q_{r}\ {}_{j}({}^{r}\Gamma)_{k}\ . \tag{1.2b}$$

Interpreting the operator ${}_{P>}O^{<Q\ \ Q>}\bar{O}_{<P}$ in an analogous manner we find the relation

$$\partial_{k}\ {}^{i}q^{j} = (\Gamma_{r})^{i}_{k}\ {}^{r}q^{j} + {}^{i}q^{r}\ {}^{j}({}_{r}\Gamma)_{k}\ . \tag{1.2c}$$

It is clear that in our discussion both points P and Q played an equivalent role and that by interchanging O and $\bar{O}$ we would obtain the relations (1.2) for the point Q . Moreover, the discussion being valid for every point $P \in \Omega$ the results are valid for all points $P \in \Omega$. We postpone to Chapter II the discussion of the relations (1.2), especially the problem of determining the coefficients of connection with the help of the fundamental tensor components.

It is clear how to generalize the notion of parallel displacement for tensors by applying the parallel displacement operators (1.1) on all basis vectors and by retaining only the first order terms in parameter differentials to obtain the parallelly displaced tensor from one point to the other.

We make another approach to the described one-to-one correspondence, i. e. parallel displacement. We introduce a correspondence between a field quantity $A(P) \in X(P)$ and a field quantity of the same type $A(P)_Q \in X(Q)$,(the points being $P(x^k)$ and $Q(x^k + dx^k)$) , with the following properties:

1) the correspondence should be linear in the sense that the following relations exist

$$[A(P) \pm B(P)]_Q = A(P)_Q \pm B(P)_Q \; , \tag{1. 3a}$$

$$[A(P) \cdot C(P)]_Q = A(P)_Q \cdot C(P)_Q \; ,$$

A, B and C being field quantities, A and B of the same type.

2) The difference of the field quantity $A(Q)$ and of $A(P)_Q$ should be a quantity of the same type of the first order in the parameter differentials

$$A(P)_Q = A(Q) - \Delta A(Q) = A(Q) - dx^k \nabla_k A(Q) \; . \tag{1. 3b}$$

We call the quantity $\Delta A(Q)$ the absolute differential and the quantity $\nabla_k A(Q)$ the $k-th$ absolute derivative of the quantity $A(Q)$.

3) The absolute differentials of the basis vectors are assumed to have these explicit forms :

$$\Delta\, {}^{i}e = dx^k\, \nabla_k\, {}^{i}e = dx^k\, {}^{i}({}_{j}\Gamma)_k\, {}^{j}e\ ,$$

$$\Delta\, {}_{i}e = dx^k\, \nabla_k\, {}_{i}e = dx^k\, {}_{i}({}^{j}\Gamma)_k\, {}_{j}e\ ,$$

$$\Delta\, e^i = dx^k\, \nabla_k\, e^i = dx^k\, (\Gamma_j)^i_k\, e^j\ , \tag{1.3c}$$

$$\Delta\, e_i = dx^k\, \nabla_k\, e_i = dx^k\, (\Gamma^j)_{ki}\, e_j\ .$$

4) The scalar quantity $f(P)_Q$ should be ideniical with the scalar quantity $f\ (P)$

$$f(P)_Q \equiv f(P)\ . \tag{1.3d}$$

From 4) and 2) the immediate consequence is that the absolute differential (derivative) of a scalar quantity should be identical with its ordinary differential (partial derivative)

$$\Delta f \equiv df\ , \qquad \nabla_k f \equiv \partial_k f\ . \tag{1.4}$$

From 2) it is clear that the parallelly displaced quantity $A\ (P)_Q$ is completely determined if the absolute differential $\Delta A(Q)$ is known. We therefore deduce the rules for the absolute differential (derivative) of a sum (difference) and of a product of field quantities, which together with (1. 3c) and (1. 3d) will allow to determine the absolute differential (derivative) of every field quantity. From 1) and 2) we immediately find

$$\Delta\,(A(Q) \pm B(Q)) = A(Q) \pm B(Q) - \left[A(P) \pm B(P)\right]_Q = \Delta\, A(Q) \pm \Delta\, B(Q) \tag{1.5a}$$

Further, we have

$$A(Q) \cdot C(Q) - \left[A(P) \cdot C(P)\right]_Q = \Delta A(Q) \cdot C(Q) + A(Q) \cdot \Delta C(Q) - \Delta A(Q) \cdot \Delta C(Q)$$

wherefrom for the first order terms in the parameter differentials the product rule follows

$$\Delta\,(A(Q) \cdot C(Q)) = \Delta\, A(Q) \cdot C(Q) + A(Q) \cdot \Delta\, C(Q)\,. \tag{1.5b}$$

The rules of the same structure (1. 5) are valid for the absolute derivatives, too.

In 3) we now take the same functions Γ as in (1. 1b). Then applying (1. 1) and (1. 3) we find

$$\begin{aligned} {}_{Q>}\bar{O}_{<P}\, {}^{i}e\,(P) &= {}^{i}e\,(P)_Q = {}^{i}e\,(Q) - \Delta\, {}^{i}e\,(Q)\,, \\ {}_{Q>}\bar{O}^{<P}\, {}_{i}e\,(P) &= {}_{i}e\,(P)_Q = {}_{i}e\,(Q) - \Delta\, {}_{i}e\,(Q)\,, \\ e^{i}\,(P)\, {}_{P>}O^{<Q} &= e^{i}\,(P)_Q = e^{i}(Q) - \Delta\, e^{i}(Q)\,, \\ e_{i}\,(P)\, {}^{P>}O_{<Q} &= e_{i}\,(P)_Q = e_{i}(Q) - \Delta\, e_{i}(Q)\,. \end{aligned} \tag{1.6}$$

Thus, with the same functions Γ in (1. 1b) and (1. 3c), the relations (1. 6) show the identity of the parallelly displaced basis vectors with the help of parallel displacement operators (1. 1) and the vectors (1. 3b) determined for basis vectors.

Moreover, from the above discussion it follows that both approaches lead to identical results regarding the parallelly displaced quantities determined with the help of the parallel displacement operators (1. 1) and the corresponding quantities determined with the help of the absolute differentials (1. 3). The correspondence satisfying the conditions 1), 2), 3) and 4) is indeed the parallel displacement produced by the operators (1. 1) satisfying the conditions a), b) and c). For a vector field we find, for instance

$$a_{<}(P)_{Q} = \left[a^{i}(P)\, e_{i}(P)\right]_{Q} = a^{i}(P)_{Q}\, e_{i}(P)_{Q} = a^{i}_{\cdot}(P)\, e_{i}(P)^{P>}\, O_{<Q} = a_{<}(P)^{P>} O_{<Q} \, .$$

This was the reason that from the beginning we used the same symbol $A\,(P)_{Q}$ for the corresponding quantity in the second approach, which is, indeed, identical with the parallelly displaced quantity in the first approach.

In the following we will develop in detail the second approach, where the notion and techniques of calculation of the absolute differential (derivative) are of basic importance.

With the relations (1. 5), (1. 3c) and (1. 4) the absolute differential (derivative) (1. 3b) of any vector and tensor is completely determined. For example, for a vector field we have

$$\Delta\,{}^{>}a = \Delta({}_{i}a\,{}^{i}e) = (\Delta\,{}_{i}a)\,{}^{i}e + {}_{i}a\,\Delta\,{}^{i}e\ .$$

Now, for a fixed index the quantity ${}_{i}a\,(x^k)$ is a scalar field quantity and (1. 4) holds. On the other hand the absolute differentials of the basis vectors are given by (1. 3c). We can therefore write

$$\Delta\,{}^{>}a = \nabla_k\,{}^{>}a\,dx^k = [\partial_k\,{}_{i}a + {}_{r}a\,{}^{r}({}_{i}\Gamma)_k]\,dx^k\,{}^{i}e =$$

$$= {}_{i}a_{|k}\,dx^k\,{}^{i}e = D_k\,{}_{i}a\,dx^k\,{}^{i}e = (D\,{}_{i}a)\,{}^{i}e = {}^{>}(D\,a)\ , \tag{1.7a}$$

where the new introduced symbols are defined implicitly;

${}_{i}a_{|k}$, $D_k\,{}_{i}a$ "the covariant derivative" and $D\,{}_{i}a =$
$dx^k\ D_k\ {}_{i}a$ the "covariant differential" of the ${}_{i}a$ vector component; the terminology used here will be explained later.

In an analogous manner we determine the absolute differential (derivative) for the other three vector forms

$$\Delta\,{}_{>}a = \nabla_k\,{}_{>}a\,dx^k = [\partial_k\,{}^{i}a + {}^{r}a\,{}_{r}({}^{i}\Gamma)_k]\,dx^k\,{}_{i}e =$$

$$= {}^{i}a_{|k}\,dx^k\,{}_{i}e = D_k\,{}^{i}a\,dx^k\,{}_{i}e = (D\,{}^{i}a)\,{}_{i}e = {}_{>}(Da)\ ,$$

$$\Lambda a^{<} = \nabla_k a^{<} dx^k = [\partial_k a_i + (\Gamma_i)^r_k a_r] dx^k e^i =$$

$$= a_{i|k} dx^k e^i = D_k a_i dx^k e^i = (D a_i) e^i = (D a)^{<} ,$$

(1.7b)

$$\Delta a_{<} = \nabla_k a_{<} dx^k = [\partial_k a^i + (\Gamma^i)_{kr} a^r] dx^k e_i =$$

$$= a^i{}_{|k} dx^k e_i = D_k a^i dx^k e_i = (D a^i) e_i = (D a)_{<} .$$

In a similar manner the absolute differential (derivative) (1. 3b) of every tensor ([1] , Chapter V) can be easily obtained in a compact form with the help of the product rule (1. 5). We show this by the example of the tensor ([1] , (5. 13))

$$\Delta({}^{>}{}_{>}T^{<}{}_{<<}) = (\Delta^{>})_{>} T^{<}{}_{<<} + {}^{>}(\Delta_{>})T^{<}{}_{<<} + {}^{>}{}_{>}(\Delta T)^{<}{}_{<<} \quad (1.8a)$$

$$+ {}^{>}{}_{>}T(\Delta^{<})_{<<} + {}^{>}{}_{>}T^{<}(\Delta_{<})_{<} + {}^{>}{}_{>}T^{<}{}_{<}(\Delta_{<}) .$$

With the abbreviations

$$(\Delta^{>}) = {}^{(>)} , \quad (\Delta_{>}) = (>) \quad etc. \quad (\Delta T) = (T)$$

for the absolute differentials of the basis vectors and tensor components in the formula (1. 8a) we obtain a much simpler expression

$$\Delta({}^{>}{}_{>}T^{<}{}_{<<}) = {}^{(>)}{}_{>}T^{<}{}_{<<} + {}^{>}{}_{(>)}T^{<}{}_{<<} + {}^{>}{}_{>}(T)^{<}{}_{<<} +$$

$$+ {}^{>}{}_{>}T^{(<)}{}_{<<} + {}^{>}{}_{>}T^{<}{}_{(<)<} + {}^{>}{}_{>}T^{<}{}_{<(<)} . \quad (1.8b)$$

To show the usefulness of the applied notation, we explicitly write the formula (1.8) using (1.3c) and (1.4) for the absolute differentials of the basis vectors and tensor components

$$\Delta\left({}^{>}_{>}T^{<}_{<<}\right) = {}^{i}e\,{}_{k}e\Big[{}^{\bar i}({}_{i}\Gamma)_{s\bar i}\,{}^{k}T_{\ell}^{\;pr} + {}_{\bar k}({}^{k}\Gamma)_{si}\,{}^{\bar k}T_{\ell}^{\;pr} + \frac{\partial}{\partial x^{s}}\,{}_{i}^{k}T_{\ell}^{\;pr} +$$

$$+ {}_{i}^{k}T_{\bar\ell}^{\;pr}(\Gamma_{\ell})^{\bar\ell}_{s} + {}_{i}^{k}T_{\ell}^{\;\bar p r}(\Gamma^{p})_{s\bar p} + {}_{i}^{k}T_{\ell}^{\;p\bar r}(\Gamma^{r})_{s\bar r}\Big]\,dx^{s}e^{\ell}e_{p}e_{r} =$$

$$= \nabla_{s}\left({}^{>}_{>}T^{<}_{<<}\right)dx^{s} = {}^{i}e\,{}_{k}e\left({}_{i}^{k}T_{\ell}^{\;pr}\right)_{|s}dx^{s}e^{\ell}e_{p}e_{r} =$$

$$= {}^{i}e\,{}_{k}e\,D_{s}\left({}_{i}^{k}T_{\ell}^{\;pr}\right)dx^{s}e^{\ell}e_{p}e_{r} = {}^{i}e\,{}_{k}e\,D\left({}_{i}^{k}T_{\ell}^{\;pr}\right)e^{\ell}e_{p}e_{r} = {}^{>}_{>}(DT)^{<}_{<<}. \tag{1.8c}$$

Here again we introduced the earlier mentioned symbols (1.7a) for the covariant differential and derivative of the tensor components, their meaning being implicitly defined by (1.8c).

From the definition (1.3b) it is clear that the absolute differential of a field quantity represents a quantity of the same type of the first order in coordinate differentials. This is explicitly confirmed in the results (1.4), (1:7) and (1.8). At the same time the product rule (1.5) reflects itself in the applied notation in a clear and evident way in forming the absolute differential of the field quantities.

From the definition (1.3b) and (1.3a) it is clear that the absolute differential of a field quantity should depend only on the points P and Q and on the "parallel displacement" of the quantity from one system of basis vectors in the vector space $X(P)$ to the system of corresponding basis vectors in the vector space $X(Q)$. This means that the absolute differential should be invariant on the change of the point coordinates (1), the corresponding basis vector systems in P and Q remaining unchanged. Thus, from (1.3) and (1) for the absolute differential (derivative) operator we obtain the expressions

$$\Delta \cdot = dx^k \nabla_k \cdot = dx'^r \nabla'_r \cdot$$

$$\nabla_k \cdot = \frac{\partial x'^r}{\partial x^k} \nabla'_r \cdot \;, \qquad \nabla'_r \cdot = \frac{\partial x^k}{\partial x'^r} \nabla_k \cdot \;, \tag{1.9a}$$

because the relations (1.3) and (1) have to be valid for any pair of points $P, Q \in \Omega$

The covariant and parallel displacement differential
(derivative)

To analyze the meaning of the absolute (covariant) differential, we interpret the definition (1. 3b) by introducing an auxiliary quantity $\bar{A}(Q)$ whose components at the point Q are those of $A(P)$ at P, e. g.

$$\bar{a}_<(Q) = \bar{a}^i(Q)\, e_i(Q) = a^i(P)\, e_i(Q) \quad . \tag{1.10}$$

Then we write the absolute differential (1. 7b$_3$) in the form

$$\begin{aligned} \Delta a_< &= a_<(Q) - \bar{a}_<(Q) + \bar{a}_<(Q) - a_<(P)_Q \\ &= \left[a^i(Q) - a^i(P) \right] e_i(Q) + a^i(P)\, \Delta e_i \end{aligned} \tag{1.11}$$

Taking into account the formulas (1. 3c) and (1. 7) we write (1. 11) as

$$\Delta a_< = da^i\, e_i + (\Gamma^i)_{kr}\, dx^k\, a^r\, e_i = D\, a^i\, e_i \quad , \tag{1.12}$$

where all the quantities refer to the same point. For the covariant differential we have the expression

$$D\, a^i = da^i + (\Gamma^i)_{kr}\, a^r\, dx^k = da^i - \delta\, a^i \tag{1.13a}$$

and for the covariant derivative of the component a^i

$$D_k\, a^i = a^i_{|k} = \partial_k\, a^i - \delta_k\, a^i \tag{1.13b}$$

Here we introduced the "parallel displacement differential (derivative)"

$$\delta a^i = \delta_k a^i dx^k = -(\Gamma^i)_{kr} a^r dx^k , \tag{1.14}$$

which would represent the first-order change of the component a^i for the parallelly displaced vector, for which $\Delta a_<$ and $D a^i$ would disappear or $a_< (P)_a = a_< (Q)$

In general, for tensor components the same relations (1.13) are valid between the covariant differential D , ordinary differential d and "parallel displacement differential" δ and between the respective derivatives. This could be easily seen from (1.8c)

$$DA = dA - \delta A ,$$

$$D_k A = \partial_k A - \delta_k A , \tag{1.15}$$

where A symbolizes any tensor component.

The rules for the covariant differential (derivative) have the same structure as those for the absolute differential (derivative) (1.5), because of the definitions (1.8)

$$D(A \pm B) = DA \pm DB , \quad D_k(A \pm B) = D_k A \pm D_k B , \tag{1.16}$$

$$D(A\cdot C) = (DA)\cdot C + A\cdot(DC)\ ,\ D_k(A\cdot C) = (D_k A)\cdot C + A\cdot(D_k C)$$

Here A, B and C are the tensor components, A and B being the tensor components of the same type. For the parallel displacement differential (derivative) $\delta = dx^k\,\delta_k$ the rules are of the form (1. 16) because of (1. 15) and the fact that for an ordinary differential (derivative) they have the same structure.

We mention that for the covariant differential (derivative) operator $D\ (D_k)$ we have the same conclusion (1. 9a) regarding the change of coordinate lines (e. g; (1. 8c))

$$D\,\cdot = dx^k\,D_k\cdot\ = dx'^r\,D'_r\cdot$$

$$D_k\cdot\ = \frac{\partial x'^r}{\partial x^k}\,D'_r\cdot\ ,\ D'_r\cdot\ = \frac{\partial x^k}{\partial x'^r}\,D_k\cdot \qquad (1.\ 9b)$$

Since the absolute differential of a tensor is a tensor of the same type, the covariant differentials of its components transform as tensor components of this type for a change of basis vectors. Thus, for a simultaneous change of coordinate lines and basis vectors we have the transformation law for covariant derivatives, for instance for the vector components (1.7a)

$$D'_{r\,\bar{i}}\bar{a}'\,dx'^r\,\overset{\bar{i}}{e} = {}^{>}\bar{E}_{<}(D_k\,{}_i a\,dx^k\,\overset{i}{e}) = \frac{\partial x^k}{\partial x'^r}(D_k\,{}_i a)\,dx'^r\,(e_{\bar{i}}\,\overset{i}{e})\,\overset{\bar{i}}{e}\,.$$

$$D'_{r\,\bar{i}}\bar{a}' = \frac{\partial x^k}{\partial x'^r}\,(D_k\,{}_i a)\,(e_{\bar{i}}\,\overset{i}{e}) \quad .$$

In analogy with (1.10), for a scalar quantity we have to take $\bar{f}(Q) = f(P)$. Consequently, in analogy with (1.11) the difference $f(P)_Q - \bar{f}(Q)$ represents the parallel displacement differential δf . It disappears identically because of (1.3d)

$$\delta f = dx^k\,\delta_k f = f(P)_Q - \bar{f}(Q) \equiv 0\,, \quad \delta_k f = 0\,. \qquad (1.17a)$$

Thus, the covariant differential Df of the scalar quantity f, in accordance with (1.15), has to be identical with the ordinary differential. Therefore, because of (1.4), for a scalar quantity we finally have

$$df \equiv Df \equiv \Delta f\,, \qquad \partial_k f \equiv D_k f \equiv \nabla_k f \quad . \qquad (1.17b)$$

Since the scalar product ([1], (2.11)) is by definition a scalar function of coordinates for any two vector fields $a(P)$ and $b(P)$, the parallel displacement differential (derivative) for every of its four forms has to vanish

$$\delta(a^{i}{}_{i}b) = \delta(a_{i}{}^{i}b) = \delta(a^{i}{}_{i}q_{\bar{\imath}}{}^{\bar{\imath}}b) = \delta(a_{i}{}^{i}q^{\bar{\imath}}{}_{\bar{\imath}}b) = 0. \quad (1.18)$$

Now, with the help of the product rule for the parallel displacement differential (derivative) (1.16), and using the explicit expressions for the parallel displacement differentials for the vector and tensor components, from (1.15), (1.7), (1.8) and the scheme ([1], (1.6)) we find again that the following relations exist (cf. (1.2))

$$(\Gamma^{\bar{\imath}})_{kr} + {}^{\bar{\imath}}({}_{r}\Gamma)_{k} = 0 \ ,$$

$$(\Gamma_{\bar{\imath}})^{r}_{k} + {}_{\bar{\imath}}({}^{r}\Gamma)_{k} = 0 \ ,$$

$$\delta\, {}_{i}q_{\bar{\imath}} = [(\Gamma^{r})_{ki}\, {}_{r}q_{\bar{\imath}} + {}_{\bar{\imath}}({}^{r}\Gamma)_{k}\, {}_{i}q_{r}]\, dx^{k} \ , \quad (1.19)$$

$$\delta\, {}^{i}q^{\bar{\imath}} = [(\Gamma_{r})^{i}_{k}\, {}^{r}q^{\bar{\imath}} + {}^{\bar{\imath}}({}_{r}\Gamma)_{k}\, {}^{i}q^{r}]\, dx^{k} \ .$$

The first two sets of relations (1.19) show again that of $4mn^2$ coefficients of connection (1.2), $2mn^2$ are dependent on the other $2mn^2$. The last two sets, with the help of the first two, are identical with the respective parallel displacement differentials (derivatives) of the fundamental tensor components deduced directly from (1.8) with the help of (1.15).

CHAPTER II

The coefficients of connection

The coefficients of connection Γ have been introduced through the explicit definitions of the absolute differentials of basis vectors (1. 3c) which connect the corresponding basis vectors of the vector spaces $X(P)$ and $X(Q)$, $P(x^k)$, $Q(x^k + dx^k) \in \Omega$. These coefficients, as defined in (1. 3c), depend on the chosen co-ordinate lines, i. e. on the coordinates x^k , and on the chosen systems of the corresponding basis vectors, i. e. on $\mathbf{e}(P)$ and $\mathbf{e}(Q)$.

We wish to investigate more closely the properties of the coefficients of connection and the respective relations. First, we note that four more sets of coefficients of connection can be introduced, because of the explicit definitions (1. 3c) and the relations ([1] , (2. 9)) and ([1] , (2. 10))

$$
{}^{i}({}^{r}\Gamma)_{k} = {}^{r}q^{j}\; {}^{i}({}_{j}\Gamma)_{k} \;, \quad (\Gamma^{r})^{i}_{k} = (\Gamma_{j})^{i}_{k}\; {}^{j}q^{r} \;,
$$

$$
{}_{i}({}_{r}\Gamma)_{k} = {}_{r}q_{j}\; {}_{i}({}^{j}\Gamma)_{k} \;, \quad (\Gamma_{r})_{ki} = (\Gamma^{j})_{ki}\; {}_{j}q_{r} \;. \tag{2. 1}
$$

From the relations (2. 1) it is clear that the inner index of the coefficients of connection has the vectorial character and can be raised and lowered with the help of the fundamental tensors, i.e. the valence raising and lowering operators ([1], (2.4)).

Now we investigate the transformation properties of the coefficients of connection.

For only a change of coordinate lines, from (1.9) and (1.3c) we obtain the relation for the coefficients of connection referring to unprimed and primed systems of coordinate lines e.g.

$$ {}^{i}({}_{j}\Gamma(x))_{k} = \frac{\partial x'^{r}}{\partial x^{k}}\, {}^{i}({}_{j}\Gamma'(x'))_{r} \; . \tag{2.2} $$

The transformation of the coefficients of connection for a change of the system of the corresponding basis vectors can also be investigated leaving the coordinate lines unchanged. Labelling the new corresponding basis vectors with a dash, we write e.g. (1.3c) with the help of the identity operator ${}^{>}\bar{E}_{<} = {}^{p}\bar{e}\;\bar{e}_{p}$ in the form

$$ \Delta({}^{>}\bar{E}_{<}\,{}^{i}e) = {}^{i}({}_{j}\Gamma)_{k}\, dx^{k}\, ({}^{>}\bar{E}_{<}\,{}^{j}e) \; , $$

$$ (\Delta\,{}^{p}\bar{e})(\bar{e}_{p}\,{}^{i}e) + {}^{p}\bar{e}\,[(\Delta\,\bar{e}_{p})\,{}^{i}e + \bar{e}_{p}\,\Delta\,{}^{i}e] = {}^{i}({}_{j}\Gamma)_{k}\, dx^{k}\, {}^{p}\bar{e}\,(\bar{e}_{p}\,{}^{j}e) \; . \tag{2.3a} $$

Substituting the respective expressions (1.3c) into (2.3a), for the old and new corresponding basis vectors we obtain the expression for the left hand side

$$ {}^{p}({}_{j}\bar{\Gamma})_{k}\, dx^{k}\, {}^{j}\bar{e}\, (\bar{e}_{p}\, {}^{i}e) + {}^{p}\bar{e}\, [(\bar{\Gamma}^{j})_{kp}\, (\bar{e}_{j}\, {}^{i}e) + {}^{i}({}_{j}\Gamma)_{k}\, (\bar{e}_{p}\, {}^{j}e)]\, dx^{k} \, . \qquad (2.\,3b) $$

From (2. 3a) and (2. 3b) we conclude that for the coefficients of connection in the new system the following relation must exist

$$ {}^{p}({}_{j}\bar{\Gamma})_{k} + (\bar{\Gamma}^{p})_{kj} = 0 \, . \qquad (2.\,4) $$

The dashed system being any of the systems of the corresponding basis vectors, the relation (2. 4) exists in every system of the corresponding basis vectors. By an analogous procedure the same relation (2. 4) can be derived from (1. $3c_4$), while from (1. $3c_2$) or (1. $3c_3$) another relation between the coefficients of connection can be proved

$$ {}_{p}({}^{j}\bar{\Gamma})_{k} + (\bar{\Gamma}_{p})^{j}_{k} = 0 \, . \qquad (2.\,5) $$

It should be pointed out that the relations (2. 4) and (2. 5) have already been proved as (1. 19) starting from the fact that the parallel displacement differential (derivative) of the scalar product vanishes.

In the case when both changes are simultaneous, i. e. the change of coordinate lines and the change of the corresponding basis vectors, we start from the absolute differential of a basis vector in the new corresponding system

$$\Delta\, {}^{i}\bar{e}' = \Delta({}^{>}E_{<}\, {}^{i}\bar{e}') = (\Delta\, {}^{p}e)\,(e_p\, {}^{i}\bar{e}') + {}^{p}e\, \Delta(e_p\, {}^{i}\bar{e}') \;,$$

$$\nabla'_r\, {}^{i}\bar{e}'\, dx'^r = \nabla_k\, {}^{p}e\, dx^k (e_p\, {}^{i}\bar{e}') + {}^{p}e \frac{\partial}{\partial x'^r}(e_p\, {}^{i}\bar{e}')\, dx'^r \;, \qquad (2.6)$$

where we used (1. 9) and (1. 4). Substituting the relations (1) and (1. 3c) for both systems, we finally obtain the transformation law

$${}^{i}({}_{\ell}\bar{\Gamma}')_r = {}^{p}({}_{i}\Gamma)_k\, (\bar{e}'_{\ell}\, {}^{i}e)\,(e_p\, {}^{i}\bar{e}') \frac{\partial x^k}{\partial x'^r} + (\bar{e}'_{\ell}\, {}^{p}e) \frac{\partial}{\partial x'^r}(e_p\, {}^{i}\bar{e}') \;. \qquad (2.7)$$

If we had started from $\Delta\, {}_{i}\bar{e}'$, in an analogous way we would have found the transformation law

$${}_{i}({}^{\ell}\bar{\Gamma}')_r = {}_{p}({}^{i}\Gamma)_k\, (\bar{e}'^{\ell}\, {}_{i}e)\,(e^{p}\, {}_{i}\bar{e}') \frac{\partial x^k}{\partial x'^r} + (\bar{e}'^{\ell}\, {}_{p}e) \frac{\partial}{\partial x'^r}(e^{p}\, {}_{i}\bar{e}') \qquad (2.8)$$

The transformation laws for the other two sets of connection coefficients $(\bar{\Gamma}'^{\ell})_{ri}$, and $(\bar{\Gamma}'_{\ell})^{i}_{r}$ can be obtained in an analogous way from the absolute differentials $\Delta\, \bar{e}'^{i}$ and $\Delta\, \bar{e}'_i$. Alternatively, in order to obtain the transformation laws for $(\bar{\Gamma}'^{p})_{ki}$ and $(\bar{\Gamma}'_{p})^{i}_{k}$ from (2. 7) and

(2. 8), we could apply the relations (2. 4) and (2. 5) and relations similar to

$$\frac{\partial}{\partial x'^r}\left[(\bar{e}'_l\ {}^p e)(e_p\ {}^i\bar{e}')\right] = \frac{\partial}{\partial x'^r}\ (\bar{e}'_l\ {}^i\bar{e}') = \frac{\partial}{\partial x'^r}\ {}_l\delta^i = 0\ . \quad (2.9)$$

From the results obtained we conclude that the transformation properties of coefficients of connection are different from those found for tensors ([1] Chapter V) and therefore they are field quantities of another kind.

For the coefficients of connection we obtain again four fundamental sets of relations (1.2) or (1. 19) by determining the absolute differential (derivative) of the quantities explicitly given in the scheme ([1] , (1. 6)). These quantities are now field quantities assumed to be known in the chosen coordinates and in the systems of the corresponding basis vectors. We start with the first of them

$$\Delta(e^i\ {}_j e) = (\Delta\ e^i)\ {}_j e + e^i\ (\Delta\ {}_j e) = \left[(\Gamma_r)^i_k\ {}^r\delta_j + {}_j({}^r\Gamma)_k\ {}^i\delta_r\right] dx^k = 0\ ,$$

where we used (1. 3c) and (1. 5). Since this relation has to be valid for all points $P,\ Q \in \Omega$, it follows

$$(\Gamma_j)^i_k + {}_j({}^i\Gamma)_k = 0\ . \quad (2.10a)$$

By an analogous procedure from the second relation ([1] ,

(1.6)) we obtain

$$(\Gamma^{j})_{ki} + {}^{j}({}_{i}\Gamma)_{k} = 0 \,. \qquad (2.10b)$$

The relations (2.10) were deduced earlier in different ways as (1.19) and (2.4), (2.5).

From the remaining two relations ([1] , (1.6)) for the fundamental tensor components we obtain in a similar way

$$\partial_k\, {}_{i}q_{j} = (\Gamma_{j})_{ki} + {}_{j}({}_{i}\Gamma)_{k} = (\Gamma^{r})_{ki}\, {}_{r}q_{j} + {}_{i}q_{r}\, {}_{j}({}^{r}\Gamma)_{k} \,,$$

$$\partial_k\, {}^{i}q^{j} = (\Gamma^{j})^{i}_{k} + {}^{j}({}^{i}\Gamma)_{k} = (\Gamma_{r})^{i}_{k}\, {}^{r}q^{j} + {}^{i}q^{r}\, {}^{j}({}_{r}\Gamma)_{k} \,. \qquad (2.11)$$

The relations (2.11) express a very important result derived with the help of the relations (1.15), (1.19) and (2.1). The parallel displacement differential (derivative) of the fundamental tensor components is identical with the ordinary differential(partial derivative) of these components. Consequently, the covariant differential (derivative) of the fundamental tensor components and the absolute differential (derivative) of the fundamental tensor itself vanish

$$d\, {}_{i}q_{j} = \delta\, {}_{i}q_{j} \;;\quad D\, {}_{i}q_{j} = 0 \,,\quad \Delta\, {}^{>}q^{<} = 0 \,,$$

$$d\, {}^{i}q^{j} = \delta\, {}^{i}q^{j} \;;\quad D\, {}^{i}q^{j} = 0 \,,\quad \Delta\, {}_{>}q_{<} = 0 \,. \qquad (2.12)$$

The same conclusion could be drawn by directly calculating the absolute differential (derivative) of the fundamental tensor ([1] , (2. 4)) with the help of (1. 8) and (2. 11).

Moreover, we can prove that the relations (2. 11) or (2. 12) are mutually dependent, only one of them being independent. This follows directly from the vanishing of the absolute differential (derivative) of the identity operators ([1] , (2. 7a))

$$\Delta\,({}^{>}q^{<}\;{}_{>}q_{<}) = (\Delta\,{}^{>}q^{<})\,{}_{>}q_{<} + {}^{>}q^{<}\,\Delta\,{}_{>}q_{<} = 0 \,. \tag{2. 13}$$

Assuming that one of the absolute differentials in (2. 13) vanishes, the other will also vanish because of ([1] , (2. 7a)). We can prove this dependence directly for the fundamental tensor components (2. 11). Here we have to use ([1], (2. 7b)) in the form

$$\partial_k\,({}_{i}q_{j}\;{}^{j}q^{r}) = 0\,, \qquad \partial_k\,({}^{i}q^{j}\;{}_{j}q_{r}) = 0 \,. \tag{2. 14}$$

Substituting one of the relations (2. 11) into (2. 14), the other relation (2. 11) is obtained.

We wish to investigate more closely the problem of interrelations between the coefficients of connection themselves, and between the coefficients of connection and the components of the fundamental tensors. We introduced $4mn^2$ coefficients of connection through the definition of the absolute differentials of the basis vectors (1. 3c). The relations (2. 1) contain additional $4mn^2$ coefficients of connection which are dependent on those in (1. 3c) and on the compcnents of the fundamental tensors. For the coefficients of connection introduced by (1. 3c) we found in (2.10) that $2mn^2$ coefficients of connection $(\Gamma_i)^i_k$, $^i({}_i\Gamma)_k$ or $(\Gamma^i)_{ki}$, ${}_i({}^i\Gamma)_k$ determine each other, so that only $2mn^2$ of them are independent. For these $2mn^2$ independent coefficients the respective relations in (2. 11) have to be fulfilled. The number of independent coefficients of connection is therefore reduced by the number N of independent equations represented by the relations (2. 11). Thus the number of independent coefficients of connection is equal to $2mn^2 - N$.

In further discussion we assume the coefficients of connection to be, in general, complex quantities, e. g.

$$(\Gamma_i)_{kj} = Re\ (\Gamma_i)_{kj} + iIm\ (\Gamma_i)_{kj} \ . \tag{2. 15}$$

Here the real and imaginary part are independent and therefore the number of terms in each set of coefficients of connection is doubled and equal to $2\,mn^2$. The real and imaginary part can be divided in the symmetric and antisymmetric part with respect to the basis vector indices, and for example, we can write (2. 15) in the form

$$(\Gamma_i)_{kj} = Re\,(\Gamma_i^S)_{kj} + Re\,(\Gamma_i^A)_{kj} + i\left[Im\,(\Gamma_i^S)_{kj} + Im\,(\Gamma_i^A)_{kj}\right] = \qquad (2.16)$$
$$= (\Gamma_i^S)_{kj} + (\Gamma_i^A)_{kj} \; .$$

We have to recall that the connections between the bra and ket forms resulted in four possibilities described in ([1] , Chapter III). Therefore we should carefully investigate each of Cases Aa), Ab), Ba), Bb) and determine the relations between the coefficients of connection and the fundamental tensors in each of them.

Case Aa)

From ([1] , (3. 8), (3. 9) and (1. 3c) (or (1. 7)) we conclude that two more sets of relations for the coefficients of connection exist in this case

$$^{j}({}_{i}\Gamma)_{k} = (\Gamma_i)^{j}_{k} \; , \qquad {}_{j}({}^{i}\Gamma)_{k} = (\Gamma^{i})_{kj} \; . \qquad (2.17)$$

These two sets of relations are not independent, one following from the other because of the relations (2. 10). Since the expressions (2. 17) represent only one independent set of mn^2 relations for the coefficients of connection, the number of independent coefficients will be $mn^2 - N$. The number of independent relations N in (2. 11) is $mn(n+1)/2$ because the fundamental tensors are symmetric ([1] , (3. 11), (3. 16)) in case Aa). Finally, we have at our disposal $mn(n-1)/2$ coefficients of connection occurring in an independent set of relations (2. 11), or taking into account (2. 15), their $mn(n-1)$ real and imaginary parts. Since from (2. 1) and (2. 17) we have

$$ {}_j({}_i\Gamma)_k = (\Gamma_i)_{kj} \quad , \quad {}^j({}^i\Gamma)_k = (\Gamma^i)^j_k \quad , \tag{2. 18} $$

we can write an independent set of relations (2. 11) in the form

$$ \partial_k \; {}_i q^s_j = (\Gamma_j)_{ki} + (\Gamma_i)_{kj} = 2\,(\Gamma^s_j)_{ki} \; . \tag{2. 19} $$

Formula (2. 19) clearly shows that the symmetric parts $(\Gamma^s_j)_{ki}$ of the coefficients of connection $(\Gamma_j)_{ki}$ are completely determined by the partial derivatives of the symmetric fundamental tensor components. The $mn(n-1)$ parts at our disposal are the antisymmetric parts of the coefficients of connection $(\Gamma_{j}^{A})_{ki}$. From the coefficients of connection $(\Gamma_{j})_{ki}$ thus determined all other coefficients of connection are determined with the help of the relations established for them earlier.

Case Ab)

In an analogous way as in case Aa) we conclude that for the coefficients of connection the following relations exist

$$^{i}({}_{i}\Gamma)_{k} = (\Gamma_{i})_{k}^{i*} \quad , \quad {}_{j}({}^{i}\Gamma)_{k} = (\Gamma^{i})_{kj}^{*} \; . \tag{2.20}$$

Again, these two sets of relations are not independent; accordingly, because of the relations (2. 10), one relation follows from the other. The number of independent real and imaginary parts of the coefficients of connection will be $2mn^2 - N'$, where N' is the number of independent relations for the real and imaginary parts in (2. 11). Now the fundamental tensors are of the form ([1], (3. 11), (3. 17)) in case Ab), and therefore N' is equal to $mn(n+1)/2 + mn(n-1)/2 = mn^2$. Thus, we conclude that mn^2 real and imaginary parts of the coefficients of connection occurring in the independent set of relations (2. 11) are at our disposal. Since from (2. 1) and (2. 20) we have

$$ {}_j({}_i\Gamma)_k = (\Gamma_i)^*_{kj} \quad , \quad {}^{j}({}^{i}\Gamma)_k = (\Gamma^i)^{j*}_k \quad , \tag{2.21}$$

we can write the independent set of relations (2. 11) in the form

$$\partial_k \; {}_i g_j = (\Gamma_j)_{ki} + (\Gamma_i)^*_{kj} \quad . \tag{2.22}$$

Because of ([1] , (3. 17)) and (2. 16) the relation (2. 22) can be separated into

$$\partial_k \; {}_i g_j^{1S} = 2\,Re\,(\Gamma_j^S)_{ki} \quad , \quad \partial_k \; {}_i g_j^{2A} = 2\,Im\,(\Gamma_j^A)_{ki} \quad . \tag{2.23}$$

(g^1 and g^2 are written instead of g_1 and g_2 [1]).
The relations (2. 23) reflect the symmetry properties of the real and imaginary part of the fundamental tensor components. From these relations $Re\ (\Gamma_j^S)_{ki}$ and $Im\ (\Gamma_j^A)_{ki}$ are completely determined with the partial derivatives of the real and imaginary parts of the fundamental tensor, respectively. The $mn^2 = mn(n-1)/2 + mn(n+1)/2$ parts of the coefficients of connection at our disposal are $Re\ (\Gamma_j^A)_{ki}$ and $Im\ (\Gamma_j^S)_{ki}$. They all together determine the coefficients of connection $(\Gamma_j)_{ki}$ (2. 16), all other coefficients of connection following from them with the help of the relations established for them earlier.

Case Ba)

From ([1], (3.8), (3.9)) and (1.3c) (or (1.7)) we infer that two additional sets of relations exist for the coefficients of connection

$$ {}_{j}({}^{i}\Gamma)_{k} = (\Gamma_{i})_{k}^{j} \,, \qquad {}^{j}({}_{i}\Gamma)_{k} = (\Gamma^{i})_{kj} \,. \tag{2.24} $$

Since in this case the components of the fundamental tensor form an orthogonal matrix ([1], (3.11), (3.18)) from (2.1) and (2.24) we obtain the relations

$$ {}_{j}({}_{r}\Gamma)_{k} = (\Gamma^{r})_{k}^{j} \,, \qquad {}^{j}({}^{r}\Gamma)_{k} = (\Gamma_{r})_{kj} \,. \tag{2.25} $$

A further consequence of (2.24) is that all coefficients of connection entering into (2.10) have to be antisymmetric i.e. of the form $(\Gamma_{i}^{A})_{k}^{j}$, $(\Gamma^{iA})_{kj}$, ${}_{j}({}^{i}\Gamma^{A})_{k}$ and ${}^{j}({}_{i}\Gamma^{A})_{k}$

The matrices of the fundamental tensors are orthogonal in case Ba) ([1] , (3.18)). Thus, (2.11) represent $mn(n-1)$ independent relations for $2mn\cdot(n-1)$ real and imaginary parts of the complex antisymmetric coefficients of connection $(\Gamma^{rA})_{ki}$ and ${}_{j}({}^{i}\Gamma^{A})_{k}$. From (2.11) and the relations with exchanged indices i and r , with the help of (2.14), we deduce this system of equations

$$[jik]\, {}^{i}g^{r} + [jrk]\, {}^{i}g^{i} = 0, \; i\, j = 1,\ldots,n; \; k = 1,\ldots,m \qquad (2.26)$$

where we used the abbreviation

$$[jik] \equiv {}_{j}({}_{i}\Gamma)_{k} - (\Gamma_{j})_{ki} \; . \qquad (2.27)$$

Equations (2.26), with indices i and r exchanged, are identical so that, in fact, the system (2.26) contains only $mn(n+1)/2$ equations for mn^2 complex quantities (2.27). The matrix of the fundamental tensor being nondegenerate, from (2.26) we can determine the adequately chosen $mn(n+1)/2$ quantities (2.27) in terms of the remaining $mn(n-1)/2$ ones which can be taken arbitrarily.

Thus, for the simplest case that these $mn(n-1)/2$ quantities are taken to be zero the other $mn(n+1)/2$ quantities should vanish too, and as the final result we have that all mn^2 quantities (2.27) vanish

$$[jik] \equiv 0, \; {}_{j}({}_{i}\Gamma)_{k} \equiv (\Gamma_{j})_{ki}, \; i,j = 1,\ldots,n; \; k = 1,\ldots,m. \qquad (2.28)$$

With this result (2.28) we can write (2.11) in the form

$$\partial_{k}\; {}_{i}g_{j} = 2(\Gamma_{j})_{ki} \; . \qquad (2.29)$$

Hence it is clear that the respective coefficients of connection $(\Gamma_j)_{ki}$ are completely determined by the partial derivatives of the fundamental tensor components, the other coefficients of connection following from them with the help of the relations established earlier.

Case Bb)

Here, taking into account that the matrix of the fundamental tensor is unitary ([1] , (3. 11), (3. 19)), in an analogous way as in case $Ba^{\bullet}$) we find the relations

$$ {}_j({}^i\Gamma)_k = (\Gamma_i)_k^{j*} \quad , \quad {}^j({}_i\Gamma)_k = (\Gamma^i)_{kj}^{*} \quad , \tag{2. 30}$$

$$ {}_j({}_r\Gamma)_k = (\Gamma^r)_k^{j*} \quad , \quad {}^j({}^r\Gamma)_k = (\Gamma_r)_{kj}^{*} \quad . \tag{2. 31}$$

A further consequence of (2. 30) is that all coefficients of connection entering into (2. 10), i. e. $(\Gamma_i)_k^j$, $(\Gamma^i)_{kj}$, ${}_j({}^i\Gamma)_k$, ${}^j({}_i\Gamma)_k$, must have the real part antisymmetric and the imaginary part symmetric e. g.

$$ (\Gamma_i)_k^j = Re\,(\Gamma_i^A)_k^j + i\,Im\,(\Gamma_i^S)_k^j \tag{2. 32}$$

i. e. they are antihermitian regarding the indices i and j

$$ (\Gamma_i)_k^j = -\,(\Gamma_j)_k^{i*} \tag{2. 33}$$

The matrices of the fundamental tensors are unitary in case Bb) ([1] , (3. 19)). Thus, (2. 11) represent mn^2 independent relations for $2mn^2$ real and imaginary parts of the complex antihermitian coefficients of connection $(\Gamma^r)_{ki}$, ${}_i(\overset{i}{\Gamma})_k$ (2.33).

Here, for the quantities (2.27) we deduce the following system of relations instead of (2.26)

$$[jik]\,{}^{j}q^{r} + ([jrk]\,{}^{j}q^{i})^{*} = 0 \tag{2.34a}$$

or

$$\begin{aligned} Re\{[jik]\,{}^{j}q^{r} + [jrk]\,{}^{j}q^{i}\} &= 0 \\ Im\{[jik]\,{}^{j}q^{r} - [jrk]\,{}^{j}q^{i}\} &= 0 \end{aligned} \tag{2.34b}$$

The first system (2.34b) contains $mn(n+1)/2$ equations and the second system $mn(n-1)/2$ equations for $2mn^2$ real and imaginary parts of the quantities (2.27). Here, therefore, from (2.34b) we can determine the adequately chosen $mn(n+1)/2$ and $mn(n-1)/2$ real and imaginary parts of the quantities (2.27) in terms of the remaining mn^2 ones which can be taken arbitrarily.

Thus, for the simplest case that these mn^2 parts are taken to be zero, the other mn^2 ones have to vanish too, and similarly to case Ba) we have as the final result that all mn^2 quantities (2.27) vanish (2.28). The result (2.28) leads again to (2.29) and to a solution of the problem how to express the coefficients of connection with the help of the partial derivatives of the fundamental tensor components in case Bb).

CHAPTER III

The commutator of the absolute differentials

In the domain Ω of the parameter manifold we take the points $P(x^k)$, $Q_1(x^k+d_1 x^k)$, $Q_2(x^k+d_2 x^k)$ and $Q(x^k+d_1 x^k+d_2 x^k)$ given in a system of coordinate lines, the points $P, Q_1, Q_2, \in \Omega$ being chosen arbitrarily. In each of these points we have the connected vector spaces $X(P)$, $X(Q_1)$, $X(Q_2)$ and $X(Q)$. In these vector spaces we take the systems of the corresponding basis vectors so that for a given connection, with the coefficients of connection $\Gamma(x^k)$, the corresponding basis vectors are related by (1. 6) to the absolute differentials explicitly given by (1. 3c).

The parallel displacement of a field quantity $A(P)$ is defined by (1. 3b), the absolute differential ΔA being a field quantity, linear in parameter differentials, and of the same type as $A(P)$. Now, we wish to investigate the parallel displacement of a field quantity more closely and are especially interested in if there exists any difference between the field quantity $A(P)_{Q_1Q}$ and $A(P)_{Q_2Q}$, the quantities parallelly displaced following the paths $P\rightarrow Q_1\rightarrow$ $\rightarrow Q$ and $P\rightarrow Q_2\rightarrow Q$. We denote the absolute differentials corresponding to the coordinate differentials $d_1 x^k$ and $d_2 x^k$ by the symbols Δ_1 and Δ_2, respectively. Then, for the path $P \to Q_1 \to Q$ we have from the definition (1. 3b)

$$A(Q_1) = A(P)_{Q_1} + \Delta_1 A(Q_1) , \qquad (3.1)$$

$$A(Q) = A(Q_1)_Q + \Delta_2 A(Q) .$$

Since for the parallel displacement the following relation holds

$$A(Q_1)_Q = A(P)_{Q_1 Q} + (\Delta_1 A(Q_1))_Q ,$$

$$\Delta_1 A(Q) = (\Delta_1 A(Q_1))_Q + \Delta_2 (\Delta_1 A(Q)) , \qquad (3.2)$$

from (3. 1) and (3. 2) we easily find

$$A(Q) - [A(P)_{Q_1 Q} + (\Delta_1 A(Q_1))_Q + \Delta_2 A(Q)] =$$

$$= \Delta_1 A(Q) - (\Delta_1 A(Q_1))_Q - \Delta_2 (\Delta_1 A(Q)) = 0 . \qquad (3.3)$$

For the path $P \to Q_2 \to Q$ a completely analogous relation will be found so that in accordance with (3. 2) we have

$$\Delta_1 A(Q) - (\Delta_1 A(Q_1))_Q = \Delta_2 (\Delta_1 A(Q)) ,$$

$$\Delta_2 A(Q) - (\Delta_2 A(Q_2))_Q = \Delta_1 (\Delta_2 A(Q)) \qquad (3.4)$$

The difference between the parallelly displaced field quantity for two different paths can be found from (3. 3) and (3. 4)

$$A(P)_{Q_1Q} - A(P)_{Q_2Q} = -(\Delta_1\Delta_2 - \Delta_2\Delta_1)A(Q) = -[\Delta]A(Q) = -[A(Q)], \quad (3.5)$$

where the abbreviation $[A(Q)]$ is introduced for the commutator of the absolute differentials acting on the field quantity $A(Q)$.

Because of the one-to-one correspondence of the parallelly displaced field quantities and their first-order identity (Chapter I), the relation (3.5) can be interpreted as a second-order change of the field quantity $A(P)_{Q_1Q}$ parallelly displaced around the closed path $Q \to Q_1 \to P \to Q_2 \to Q$.

We have the "absolute parallelism" for the field quantities when for any closed path the difference (3.5) vanishes

$$A(P)_{Q_1Q} - A(P)_{Q_2Q} = 0, \text{ i.e. } [A(Q)] = [\Delta]A(Q) = 0, \quad (3.6)$$

or, what is the same, that the parallelly displaced field quantity does not depend on the path between the points P and Q. The absolute parallelism is realized in a domain of the parameter manifold, as seen from (3.5), if the commutator of the absolute differentials disappears in this domain.

In Chapter I we saw that for the differential operators, linear in the coordinate differentials (the absolute differential Δ, the covariant differential D, the parallel displacement differential δ), the same rules

(1.5) are valid as for the ordinary differential d

$$\Theta(A \pm B) = \Theta A \pm \Theta B \,, \qquad \Theta = \Delta, d, D, \delta$$

$$\Theta(A \cdot C) = (\Theta A) \cdot C + A \cdot (\Theta C) \,. \tag{3.7}$$

For the corresponding derivatives the rules of the same structure (3.7) are valid, too

$$\Theta_k(A \pm B) = \Theta_k A \pm \Theta_k B \,, \qquad \Theta = dx^k \, \Theta_k \,,$$

$$\Theta_k(A \cdot C) = (\Theta_k A) \cdot C + A \cdot (\Theta_k C) \,. \quad \Theta_k = \nabla_k, D_k, \delta_k, \partial_k \tag{3.8}$$

According to the definition of the differential operators, the field quantities A, B and C are tensors or their components, which are scalar functions of the coordinates.

Now, we form the commutators of the differentials (derivatives)

$$[\Theta] = (\Theta_1 \Theta_2 - \Theta_2 \Theta_1) \,, \qquad [\Theta]_{jk} = \Theta_j \Theta_k - \Theta_k \Theta_j \,,$$

$$\Theta_1 = d_1 x^i \, \Theta_i \,, \qquad \Theta_2 = d_2 x^k \, \Theta_k \,, \tag{3.9}$$

and prove that rules analogous to (3.7) and (3.8) are valid for them, too. Indeed, from (3.7) written for Θ_1 and Θ_2

by acting on them with Θ_2 and Θ_1 , respectively, we find

$$[\Theta](A \pm B) = [\Theta] A \pm [\Theta] B ,$$
$$[\Theta](A \cdot C) = ([\Theta] A) \cdot C + A \cdot ([\Theta] C) . \tag{3.10}$$

In a completely analogous manner the relations (3. 8) yield

$$[\Theta]_{jk} (A \pm B) = [\Theta]_{jk} A \pm [\Theta]_{jk} B ,$$
$$[\Theta]_{jk} (A \cdot C) = ([\Theta]_{jk} A) \cdot C + A \cdot ([\Theta]_{jk} C) . \tag{3.11}$$

Thus, the rules of the same structure (3. 7) and (3. 8) are valid for the linear differential operators Δ, D, δ, d (3. 7) and the respective derivatives (3. 8) as well as for the commutators of these differentials (3. 10) and the commutators of the respective derivatives (3. 11).

The relations between the commutators of the differentials and the commutators of the respective derivatives are obtained from (3. 9) and (3. 7) by writing explicitly

$$\Theta_1(\Theta_2 A) = \Theta_1 [(\Theta_k A) d_2 x^k] = \Theta_i(\Theta_k A) d_2 x^k d_1 x^i + (\Theta_k A) \Theta_1 d_2 x^k ,$$
$$\Theta_2(\Theta_1 A) = \Theta_2[(\Theta_i A) d_1 x^i] = \Theta_k(\Theta_i A) d_1 x^i d_2 x^k + (\Theta_i A) \Theta_2 d_1 x^i . \tag{3.12}$$

From the relations (3. 12) we immediately conclude that the relation between the commutators of the differentials and the respective derivatives is of the form

$$[\Theta]A = ([\Theta]_{jk}A)\, d_2x^k d_1x^j + (\Theta_k A)\,(\Theta_1 d_2x^k - \Theta_2 d_1x^k). \quad (3.13)$$

For the ordinary differentials $\Theta \equiv d$ and the partial derivatives $\Theta_k \equiv \partial_k$ of the scalar functions of parameters, $A(x) \in C^N$, $N \geq 2$ the relation (3.13) is evidently fulfilled, each term vanishing separately

$$[d]\,A(x) \equiv d_1d_2A - d_2d_1A \equiv 0,\ [\partial]_{jk}A = (\partial^2_{jk} - \partial^2_{kj})A \equiv 0. \quad (3.14)$$

When $\Theta \equiv \Delta, \Theta_k \equiv \nabla_k$, we easily see that the last term of the relation (3.13) vanishes because of the relation (1.4). Thus, the commutator of the absolute differentials and the commutator of the absolute derivatives are connected as

$$[\Delta]\,A = [\nabla]_{jk}\,A\, d_2x^k d_1x^j. \quad (3.15)$$

For the covariant differential $\Theta \equiv D$ and the parallel displacement differential $\Theta \equiv \delta$ the second terms on the right-hand side of (3.13) vanish, too, because of (1.17) and (3.14). Namely, at the present moment the parameter differentials are not yet vector components in the n-dimensional vector space $X(P)$.

We go over to determine the commutators of the absolute differentials acting on scalars, vectors and, in general, on tensors.

If we have a scalar function $f(x^k) \in C^N$, $N \geq 2$, dependent on the coordinates, we easily determine

$$[\Delta] f = [f] = 0 , \tag{3.16}$$

because of the relations (1.4) and (3.14).

The commutators of the absolute differentials of the basis vectors can be found from (1.3c) and (3.10) (or (3.11)). We start with the basis vectors $\overset{i}{e}$

$$[\Delta] \overset{i}{e} = [\overset{i}{e}] - [\nabla]_{jk} \overset{i}{e}\, d_1 x^j d_2 x^k . \tag{3.17}$$

From (1.3c) we directly find

$$\nabla_j \nabla_k \overset{i}{e} = \nabla_j [{}^i({}_r\Gamma)_k \overset{r}{e}] - [\partial_j\ {}^i({}_r\Gamma)_k + {}^i({}_q\Gamma)_k\ {}^q({}_r\Gamma)_j] \overset{r}{e} , \tag{3.18}$$

the term $\nabla_k \nabla_j \overset{i}{e}$ being expressed by (3.18), where the indices j and k are exchanged. Thus, we obtain (3.17) in the form

$$[\Delta]\,\overset{i}{e} = \overset{r}{e}\,[\partial_j\,{}^i({}_r\Gamma)_k - \partial_k\,{}^i({}_r\Gamma)_j + {}^i({}_q\Gamma)_k\,{}^q({}_r\Gamma)_j - {}^i({}_q\Gamma)_j\,{}^q({}_r\Gamma)_k]\,d_1x^j d_2x^k =$$

$$= \overset{r}{e}\;{}_rR_{kj}{}^i\,d_1x^j d_2x^k \;.$$

(3. 19a)

In a completely analogous way we find the expressions for the commutators of the absolute differentials for the other three sets of basis vectors

$$[\Delta]\,{}_ie = {}_re\,[\partial_j\,{}_i({}^r\Gamma)_k - \partial_k\,{}_i({}^r\Gamma)_j + {}_i({}^q\Gamma)_k\,{}_q({}^r\Gamma)_j - {}_i({}^q\Gamma)_j\,{}_q({}^r\Gamma)_k]\,d_1x^j d_2x^k =$$

$$= {}_re\;{}^rR_{kji}\,d_1x^j d_2x^k \;,$$

$$[\Delta]\,e^i = d_2x^k d_1x^j\,[\partial_j(\Gamma_r)^i_k - \partial_k(\Gamma_r)^i_j + (\Gamma_q)^i_k(\Gamma_r)^q_j - (\Gamma_q)^i_j(\Gamma_r)^q_k]\,e^r =$$

$$= d_2x^k d_1x^j\;{}^i{}_{jk}R_r\,e^r \;,$$ (3. 19b)

$$[\Delta]\,e_i = d_2x^k d_1x^j\,[\partial_j(\Gamma^r)_{ki} - \partial_k(\Gamma^r)_{ji} + (\Gamma^q)_{ki}(\Gamma^r)_{jq} - (\Gamma^q)_{ji}(\Gamma^r)_{kq}]\,e_r =$$

$$= d_2x^k d_1x^j\;{}_{ijk}R^r\,e_r \;.$$

In the formulas (3. 19) we introduced the quantities R, with four indices as abbreviations for the expressions in the brackets composed of coefficients of connection and their partial derivatives. The pair of indices (k, j) refers to the coordinate differentials, while the remaining indices i and r of the quantities R refer to the corresponding basis vectors. From the relations (3.19) we immediately conclude that the quantities R are antisymmetric in the indices (k, j), e.g.

$$ {}_{r}R_{kj}{}^{i} = - {}_{r}R_{jk}{}^{i} . \tag{3.20}$$

With the help of the relations (2.10) another property regarding the transposition of the pairs of indices (k, j) becomes evident from (3.19)

$$ {}_{i\,jk}R^{r} = {}_{i}R_{jk}{}^{r} , \quad {}^{i}{}_{jk}R_{r} = {}^{i}R_{jkr} . \tag{3.21}$$

The conclusion (3.21) could also be deduced from the scheme ([1] , (1.6)) by applying the commutator to the first two relations, the rule (3.10) and the definitions of the quantities R (3.19) being taken into account. The vector character of the indices i and r becomes evident from the other two relations ([1] , (1.6)) if treated similarly, or directly from the relations (3.19).

Thus, we can define the linear second-order operator

$$ {}_{>}P^{<} = {}_{r}e\, {}^{r}P_{i}\, e^{i} = {}_{r}e\,({}^{r}R_{kji}\, d_1 x^{j}\, d_2 x^{k})\, e^{i} , \qquad (3.22) $$

whose action on the basis vectors is equivalent to the action of the commutator of the absolute differentials on the basis vectors (3.19)

$$ [\Delta]\, {}^{i}e = {}^{>}P_{<}\, {}^{i}e = {}^{r}e\, {}_{r}P^{i} , \qquad [\Delta]\, e^{i} = -e^{i}\, {}_{>}P^{<} = -{}^{i}P_{r}\, e^{r} , $$

$$ [\Delta]\, {}_{i}e = {}_{>}P^{<}\, {}_{i}e = {}_{r}e\, {}^{r}P_{i} , \qquad [\Delta]\, e_{i} = -e_{i}\, {}^{>}P_{<} = -{}_{i}P^{r}\, e_{r} . \qquad (3.23) $$

The transformation properties of the quantities R follow from (1) for a change of coordinate lines and from ([1], Chapter IV) for a change of the systems of the corresponding basis vectors. Thus, both changes being made simultaneously, the transformation law is given by the relation

$$ {}_{>}\bar{P}'^{<} = {}_{>}\bar{E}^{<}\, {}_{r}e\,\left({}^{r}R_{kji}\, \frac{\partial x^{j}}{\partial x'^{\bar{j}}}\, \frac{\partial x^{k}}{\partial x'^{\bar{k}}}\, d_1 x'^{\bar{j}}\, d_2 x'^{\bar{k}}\right) e^{i}\, {}_{>}\bar{E}^{<} . \qquad (3.24) $$

For the components of the transformed operator (3.22) it therefore exists

$$ {}^{\bar r}P'_{\bar i} = {}^{\bar r}R'_{\bar k \bar j \bar i} \, d_1 x'^{\bar j} \, d_2 x'^{\bar k} = $$

$$ = (e^{\bar r} {}_r e) \, {}^r R_{kji} \frac{\partial x^j}{\partial x'^{\bar j}} \frac{\partial x^k}{\partial x'^{\bar k}} (e^i {}_{\bar i} e) \, d_1 x'^{\bar j} \, d_2 x'^{\bar k} \, . \qquad (3.25) $$

The transformation law for a separate change of coordinate lines or for a separate change of the systems of the corresponding basis vectors are trivial cases of (3.24) or (3.25).

After the expressions for the commutator of the absolute differentials of the scalar function (3.16) and of the basis vectors (3.19) have been established it is easy to obtain the result of the action of the commutators for vectors and, in general, for tensors. We use the product rule (3.10) for the commutator of the absolute differentials of a vector to obtain e.g.

$$ [\Delta]\, {}^{>}a = [{}^{>}a] = {}^{[>]}a = [\Delta]({}_i a \, {}^i e) = {}_i a \, [\Delta]\, {}^i e = {}_i a \, {}^{>}P_{<} \, {}^i e = $$

$$ = {}_r P^i \, {}_i a \, {}^r e = {}_r R_{kj}{}^i \, {}_i a \, d_1 x^j d_2 x^k \, {}^r e \, . $$

(3.26a)

Here we have used the expressions (3.19) and (3.23) and the fact that the commutator of the absolute differentials of the vector components ${}_i a \; (x^k)$ vanishes because of (3.16).

For the other three forms of vectors we have analogously

$$[\Delta]\,{}_{>}a = [{}_{>}a] = {}_{[>]}a = [\Delta]({}^{i}a\,{}_{i}e) = {}^{i}a\,[\Delta]\,{}_{i}e = {}^{i}a\,{}_{>}P^{<}\,{}_{i}e =$$

$$= {}^{r}P_{i}\,{}^{i}a\,{}_{r}e = {}^{r}R_{kji}\,{}^{i}a\,d_1 x^j d_2 x^k\,{}_{r}e\,,$$

$$[\Delta]\,a_{<} = [a_{<}] = a_{[<]} = [\Delta](a^i e_i) = a^i[\Delta]\,e_i = -a^i e_i\,{}^{>}P_{<} =$$

$$= -a^i\,{}_{i}P^{r} e_r = -a^i\,{}_{i}R_{kj}{}^{r}\,d_1 x^j d_2 x^k e_r\,, \qquad (3.26b)$$

$$[\Delta]\,a^{<} = [a^{<}] = a^{[<]} = [\Delta](a_i e^i) = a_i[\Delta]\,e^i = -a_i e^i\,{}_{>}P^{<} =$$

$$= -a_i\,{}^{i}P_{r}\,e^r = -a_i\,{}^{i}R_{kjr}\,d_1 x^j d_2 x^k e^r\,.$$

To illustrate the procedure for obtaining the commutator of the absolute differentials for tensors, we take, for example, ([1] , (5.13))

$$[\Delta]({}^{>}{}_{>}T^{<}{}_{<<}) = [{}^{>}{}_{>}T^{<}{}_{<<}] = [\Delta]({}^{i}e_{k}e\,{}_{i}{}^{k}T_{\ell}{}^{pr}\,e^{\ell} e_p e_r)\,. \qquad (3.27a)$$

The product rule (3. 10) and the fact that the commutators of the absolute differentials of the tensor components vanish by (3. 16) make it possible to write (3. 27a) in a very compact form

$$[{}^{>}{}_{>}T^{<}{}_{<<}] = {}^{[>]}{}_{>}T^{<}{}_{<<} + {}^{>}{}_{[>]}T^{<}{}_{<<} + {}^{>}{}_{>}T^{[<]}{}_{<<} +$$

$$+ {}^{>}{}_{>}T^{<}{}_{[<]<} + {}^{>}{}_{>}T^{<}{}_{<[<]} \, . \qquad (3.27b)$$

The valences indicated in the bracket mean that in the explicitly written formula the commutators (3. 23) of the corresponding basis vectors have to be substituted

$$[{}^{>}{}_{>}T^{<}{}_{<<}] = ({}^{>}P_{<}{}^{>})\,{}_{>}T^{<}{}_{<<} + {}^{>}({}_{>}P^{<}{}_{>})\,T^{<}{}_{<<} -$$

$$- {}^{>}{}_{>}T({}^{<}{}_{>}P^{<})_{<<} - {}^{>}{}_{>}T^{<}({}_{<}{}^{>}P_{<})_{<} - {}^{>}{}_{>}T^{<}{}_{<}({}_{<}{}^{>}P_{<}) \, . \qquad (3.27c)$$

Here, again, we have to substitute the explicit expressions for the commutators (3. 23), in terms of the operator components (3. 22), or (3. 19) in terms of the quantities R , e. g.

$$
\begin{aligned}
[{}^{>}{}_{>}T^{<}{}_{<<}] = {}^{i}e\,{}_{k}e\,[\,{}_{i}R_{st}{}^{\bar{i}}{}_{\bar{i}}{}^{k}T_{\ell}{}^{pr} + {}^{k}R_{st\,\bar{k}}\,{}_{i}{}^{\bar{k}}T_{\ell}{}^{pr} + \\
+ {}_{i}{}^{k}T_{\bar{\ell}}{}^{pr}\,{}^{\bar{\ell}}{}_{ts}R_{\ell} + {}_{i}{}^{k}T_{\ell}{}^{\bar{p}r}\,{}_{\bar{p}ts}R^{p} + {}_{i}{}^{k}T_{\ell}{}^{p\bar{r}}\,{}_{\bar{r}ts}R^{r}\,] \times \\
\times\, d_1 x^t\, d_2 x^s\; e^{\ell}\, e_p\, e_r \quad .
\end{aligned}
\tag{3.27d}
$$

We end the discussion with the remark that in the second part of the paper the very important case of equal dimensions of the parameter manifold and the vector spaces will be treated in detail.

R e f e r e n c e s

[1] Z. Janković : A contribution to the vector and tensor algebra

A CONTRIBUTION TO THE VECTOR AND TENSOR ANALYSIS II

-In a previous paper [2] we developed the basic features of the vector and tensor analysis under the assumption that there is an n -dimensional vector space at every point of an m -dimensional parameter manifold. Here we make the final step to connect the vector spaces $X(P)$, $P \in \Omega$ with the parameter manifold as far as possible. Such a manifold with a structure determined by the fundamental tensor and by the connection coefficients may be called a space. We therefore require the dimension m of the parameter manifold and the dimension n of the vector spaces $X(P)$, $P \in \Omega$ to be equal, $m = n$. Naturally, the results obtained for the case m not necessarily equal to n [2] , when adapted to the case $m = n$, remain conserved, while some new results and relations, characteristic of the case $m = n$, will be derived.

In Chapter I we define the displacement vector whose components are expressed by coordinate differentials in the system of corresponding basis vectors which are "tangent" to the coordinate lines at every point of the manifold. The notions of distance, metric tensor, and metric space are explained. The transformation coefficients are expressed with the help of the partial derivatives of coordinates and explicit transformation formulas for vectors and tensors are given. The connection is studied in Chapter II, i.e. the transformation properties of the coefficients of connection, the torsion tensor and the problem of determining coefficients of connection in terms of fundamental tensor components. In Chapter III the fundamental properties of the curvature tensor are established and analyzed in detail with the help of absolute derivatives and their commutators of the tangent basis vectors. Some remarks are added in Chapter IV.

CHAPTER I

The metric space

First, we concentrate our attention on the definition of the "displacement vector" $\boldsymbol{dx}$ in the vector space $X(P)$, $P \in \Omega$ which should connect the "distance" of the points $P(x^k)$ and $Q(x^k + dx^k)$ with coordinate (parameter) differentials in the chosen system of coordinate lines. Symbolically, in a system of corresponding basis vectors $\boldsymbol{e}(P) \in X(P)$ we write four forms of the displacement vector

$$(dx)^{<} = e^{i}(dx)_{i} \ , \ (dx)_{<} = e_{i}(dx)^{i} \ ,$$

$$^{>}(dx) = {}_{i}(dx)\,{}^{i}e \ , \ {}_{>}(dx) = {}^{i}(dx)\,{}_{i}e \ . \tag{1.1}$$

Afterwards we define the "distance" $\overline{PQ} = ds$ as the positive square root of the scalar product of the displacement vector multiplied by itself

$$\overline{PQ}^{2} = ds^{2} = (dx)_{<}\,{}^{>}(dx) \ . \tag{1.2}$$

Here, four equivalent expressions for the scalar product ([1], (2.11)) on the right-hand side of (1.2) can be used.

At the end we take into account that the coordinates (parameters) are assumed to be real. We can therefore identify the vector components (1.1) with the coordinate differentials in the following manners :

	$(dx)^{i} =$	$(dx)_{i} =$	${}^{i}(dx) =$	${}_{i}(dx) =$	$(dx)_{<}\,{}^{>}(dx) =$
α) A	dx^{i}	$dx^{j}\,{}_{j}g_{i}$	dx^{i}	${}_{i}g_{j}\,dx^{j}$	${}_{i}g_{j}\,dx^{i}\,dx^{j}$
B	dx^{i}	$dx^{j}\,{}_{j}g_{i}$	${}^{i}g^{j}\,dx^{j}$	dx^{i}	$dx^{i}\,dx^{i}$
β) A	$dx^{j}\,{}^{j}g^{i}$	dx^{i}	${}^{i}g^{j}\,dx^{j}$	dx^{i}	${}^{j}g^{i}\,dx^{i}\,dx^{j}$
B	$dx^{j}\,{}^{j}g^{i}$	dx^{i}	dx^{i}	${}_{i}g_{j}\,dx^{j}$	$dx^{i}\,dx^{i}$

(1.3)

The identification for $^{i}(dx)$ and $_{i}(dx)$ is given by $\alpha A, \beta B$ and αB, βA, respectively.

We see from (1.3) that the fundamental tensors appear in the expression for the distance(1.2) in cases αA, βA only. In these cases only the symmetric parts of the fundamental tensors play a role, while the difference between the expression (1.3 αA) and (1.3 βA) is derived from the different assignment of the displacement vector components and the coordinate differentials. Thus, both cases being equivalent, it is sufficient to discuss one of them, say αA; the other case follows from it with the help of ([1] , (2.10)). We see from (1.3) that in cases αB and βB the square of distance (1.2), being equal to the positive definite sum of squares of the coordinate differentials, contains no fundamental tensor components.

In the following we discuss a more general case αA under the assumption that the square of distance (1.2)

$$ds^2 = {}_{i}g_{j}\, dx^i\, dx^j \tag{1.4}$$

is real. It follows that the symmetric part of the fundamental tensor, the so-called metric tensor, has to be a real quantity, too. Then, recalling the results([1] , (3.16). (3.17)) we conclude that the fundamental tensor can have the following forms only

$$\text{Aa)} \quad g = g^{1S}, \qquad \text{Ab)} \quad g = g^{1S} + i\, g^{2A} \qquad (1.5)$$

(g^1 and g^2 are written instead of g_1 and g_2 [1]).

In case (1. 3 αA) the four forms of the displacement vector (1. 1) are explicitly expressed by

$$(dx)^{<} = dx^{j}\, {}_{j}g_{i}\, e^{i}, \quad (dx)_{<} = dx^{i}\, e_{i} \ ,$$

$$^{>}(dx) = {}^{i}e\, {}_{i}g_{j}\, dx^{j}, \quad {}_{>}(dx) = {}_{i}e\, dx^{i} \ . \qquad (1.6)$$

For a particular case where the point Q lies on a coordinate line passing through the point P, only the differential of that coordinate is different from zero, for instance $Q_1(x^1+dx^1, x^2, \ldots, x^n)$. In this case for the displacement vector (1. 6) and the distance (1. 4) we find the expressions

$$(dx)_1^{<} = dx^1\, {}_1g_i\, e^i, \quad (dx)_{1<} = e_1\, dx^1 \ ,$$

$$^{>}(dx)_1 = {}^{i}e\, {}_{i}g_1\, dx^1, \quad {}_{>}(dx)_1 = {}_1e\, dx^1 \ ,$$

$$ds_1^2 = \overline{PQ}_1^{\,2} = {}_1g_1\, (dx^1)^2 \ . \qquad (1.7)$$

Hence the displacement vector $(dx)_1$ could be said to be

"tangent" to the first coordinate line at the point P , and at the same time the corresponding basis vectors could be said to be "tangent" to this line, too. Thus, we see that for every point $P(x^k) \in \Omega$ a one-to-one correspondence between the coordinate lines and the corresponding basis vectors has been established by the relations (1.6) in the sense that the basis vectors have to be tangent basis vectors to the chosen coordinate lines. Therefore, the change of coordinate lines and the change of the corresponding basis vectors are no longer independent one of the other as in [2] , but become a simultaneous change of the system of coordinate lines and the corresponding tangent basis vectors.

The parameter manifold with the displacement vector and the distance defined by (1.6) for all $P, Q \in \Omega$ in any system of coordinate lines ([2] , (1)) and corresponding basis vectors in the vector space $X(P)$ is called the n -dimensional metric space. The metric tensor (1.5) is the symmetric part of the fundamental tensor.

In the metric space we could start from another system of coordinate lines ([2] , (1)) and corresponding tangent basis vectors $\boldsymbol{e}'$, the points P and Q being labelled by $P(x'^i)$ and $Q(x'^i + dx'^i)$, respectively. The displacement vector (1.6) in both systems of coordinate lines and corresponding tangent basis vectors is given by the respective expressions

$$(dx)_< = e_k\, dx^k = e'_j\, dx'^j = (e_k\, {}^j e')\, dx^k\, e'_j = (e'_j\, {}^k e)\, dx'^j\, e_k\, , \tag{1.8}$$

$$_>(dx) = {}_k e\, dx^k = {}_j e'\, dx'^j = (e'^j\, {}_k e)\, dx^k\, {}_j e' = (e^k\, {}_j e')\, dx'^j\, {}_k e\, ,$$

where the last expressions have been obtained by applying the identity operators ([1] , (2.1)). With the help of ([2] , (1)), from (1.8) we obtain the transformation coefficients ([1] , Chapter IV) expressed as functions of coordinates

$$\frac{\partial x^k}{\partial x'^j} = (e^k\, {}_j e') = (e'_j\, {}^k e)\, , \qquad \frac{\partial x'^j}{\partial x^k} = (e'^j\, {}_k e) = (e_k\, {}^j e')\, . \tag{1.9}$$

From these expressions it is immediately clear that the relations ([1] , (4.2)) are fulfilled and that the group property ([1] , (4.6)) is realized. An analogous treatment of the other two forms (1.6) of the displacement vector leads to the transformation property of the fundamental tensor components

$${}_i g'_j = {}_{\bar{i}} g_{\bar{j}}\, \frac{\partial x^{\bar{i}}}{\partial x'^i}\, \frac{\partial x^{\bar{j}}}{\partial x'^j}\, , \tag{1.10}$$

in accordance with the general transformation law for the tensor components (1.11).

The invariance of the distance(1.4) with respect to the coordinate transformations ([2] , (1)) is directly proved with the help of (1.10) and ([2] , (1)). Alternatively, the invariance of the expression (1.2) and consequently of (1.4), follows from the invariance of the scalar product with respect to the basis vectors transformation ([1] , (4.5)).

The names "covariant" and "contravariant" can be introduced as a convention for valences and vector and tensor components which are transformed from the original system into the primed system of coordinate lines(or tangent basis vectors) with the transformation coefficients

$\frac{\partial x^k}{\partial x'^i}$ and $\frac{\partial x'^i}{\partial x^k}$, respectively. Also, the names "up" and "down" ([1] , Chapter I) could be exchanged with "covariant" and "contravariant" when referring to valences; opposite terms have to be used for component indices.

In the metric space it is easy to express the transformation formulas ([1] , Chapter IV) with the help of (1.9), e.g.

([1], (4.1)) $\quad e^i = e'^i \frac{\partial x^i}{\partial x'^i} \; , \quad e'^i = e^i \frac{\partial x'^i}{\partial x^i} \; ,$

([1], (4.3)) $$a_i = a'_{\bar{i}} \frac{\partial x'^{\bar{i}}}{\partial x^i} , \quad a'_{\bar{i}} = a_i \frac{\partial x^i}{\partial x'^{\bar{i}}} ,$$

(1.11)

([1], (5.12)) $${}_{\bar{i}}^{\bar{k}}T'^{\bar{p}\bar{r}}_{\bar{l}} = {}_{i}^{k}T^{pr}_{l} \frac{\partial x^i}{\partial x'^{\bar{i}}} \frac{\partial x'^{\bar{k}}}{\partial x^k} \frac{\partial x^l}{\partial x'^{\bar{l}}} \frac{\partial x'^{\bar{p}}}{\partial x^p} \frac{\partial x'^{\bar{r}}}{\partial x^r} .$$

A general rule for the transformation of tensor components can be easily deduced from the third expression (1.11).

We are now able to explain the term "covariant derivative" with the help of the transformation properties of this quantity. We explain it by the example ([2], (1.7b))

$$\Delta \underset{>}{a} = (D_k \overset{i}{a}) (dx)_< \overset{k}{e} \underset{i}{e} = \overset{i}{a}_{|k} (dx)_< \overset{k}{e} \underset{i}{e} , \tag{1.12a}$$

where we used (1.8) to represent dx^k. The absolute differential (1.12a), as a vector, can be represented in the primed system with the help of the identity operator

$$(\Delta \underset{>}{a})' = \overset{i}{a}_{|k} (dx)_{<'} \overset{k}{e}' (e'_r \overset{k}{e}) (e'^p \underset{i}{e}) \underset{p}{e}' = \overset{p}{a}'_{|r} (dx)_{<'} \overset{r}{e}' \underset{p}{e}' .$$

(1.12b)

Hence the transformation rule for the covariant derivative is (cf. [2], Chapter I) :

$$ {}^{p}a'|_{r} = {}^{i}a|_{k} \, (e'_{r} \, {}^{k}e) \, (e'^{p} \, {}_{i}e) = {}^{i}a|_{k} \, \frac{\partial x'^{p}}{\partial x^{i}} \, \frac{\partial x^{k}}{\partial x'^{r}} \quad . \tag{1.13} $$

Formulas (1.13) and (1.11) show that the covariant derivatives of vector components are transformed as components of a second-rank tensor, the index r (and k) being a covariant index. For the absolute differentials of three other forms of vectors ([2] , (1.7)) and of tensors ([2] , (1.8)) an analogous conclusion can be drawn.

CHAPTER II

The connection in the metric space

The connection of the vector spaces $X(P)$ and $X(Q)$, $P, Q \in \Omega$ was introduced in a previous paper ([2] , Chapter II), where the dimension of the parameter manifold and the dimension of the vector spaces were in general different, $n \neq m$. Since for the metric space these dimensions are equal, coefficients of connection can be divided in the symmetric and antisymmetric parts with respect to the outer indices, analogously to ([2] , (2.16)), e.g.

$$(\Gamma^{j})_{ki} = (\Gamma^{j})^{S}_{ki} + (\Gamma^{j})^{A}_{ki} =$$

$$= Re\,(\Gamma^{j})^{S}_{ki} + Re\,(\Gamma^{j})^{A}_{ki} + i\left[Im\,(\Gamma^{j})^{S}_{ki} + Im\,(\Gamma^{j})^{A}_{ki}\right] \tag{2.1}$$

The transformation properties of the coefficients of connection follow from ([2] , (2.7), (2.8)) and (1.9)

$$\begin{aligned} {}^{i}_{\ell}(\Gamma')_{r} &= {}^{p}_{j}(\Gamma)_{k} \frac{\partial x^{j}}{\partial x'^{\ell}} \frac{\partial x'^{i}}{\partial x^{p}} \frac{\partial x^{k}}{\partial x'^{r}} + \frac{\partial x^{p}}{\partial x'^{\ell}} \frac{\partial}{\partial x'^{r}} \left(\frac{\partial x'^{i}}{\partial x^{p}} \right), \\ {}_{i}({}^{\ell}\Gamma')_{r} &= {}_{p}({}^{j}\Gamma)_{r} \frac{\partial x'^{\ell}}{\partial x^{j}} \frac{\partial x^{p}}{\partial x'^{i}} \frac{\partial x^{k}}{\partial x'^{r}} + \frac{\partial x'^{\ell}}{\partial x^{p}} \frac{\partial}{\partial x'^{r}} \left(\frac{\partial x^{p}}{\partial x'^{i}} \right), \end{aligned} \tag{2.2}$$

and because of ([2] , (2.1), (2.10), (2.17), (2.20)), they implicitly represent the transformation rules for all coefficients of connection in case A.

It is evident from (2.2) that the coefficients of connection are not components of a third-rank tensor because of the second member on the right-hand side.

However, the first members on the right - hand side are transformed as components of a third-rank tensor (1. 11). From the second relation (2. 2) we therefore conclude that the following transformation formula is valid

$$(\Gamma'^{\ell})_{ri} - (\Gamma'^{\ell})_{ir} = [(\Gamma^{j})_{kp} - (\Gamma^{j})_{pk}] \frac{\partial x'^{\ell}}{\partial x^{j}} \frac{\partial x^{p}}{\partial x'^{i}} \frac{\partial x^{k}}{\partial x'^{r}} . \tag{2. 3}$$

Thus, we can define a third-rank tensor, the so-called torsion tensor whose explicit form is suggested by (2. 3)

$$_{>}S^{<<} = {}_{j}e\,[(\Gamma^{j})_{kp} - (\Gamma^{j})_{pk}]\, e^{k} e^{p} = {}_{j}e\; 2\,(\Gamma^{j})^{A}_{kp}\, e^{k} e^{p} . \tag{2. 4}$$

It is interesting to note that the property (2. 3) of the coefficients of connection is implicitly contained in the commutability of the partial derivatives of coordinates ([2] , (1)), i. e.

$$\frac{\partial^2 x'^{j}}{\partial x^{i} \partial x^{m}} = \frac{\partial^2 x'^{j}}{\partial x^{m} \partial x^{i}} , \quad \frac{\partial^2 x^{j}}{\partial x'^{i} \partial x'^{m}} = \frac{\partial^2 x^{j}}{\partial x'^{m} \partial x'^{i}} . \tag{2. 5}$$

With the help of the relations (1. 9) and ([2] , (1. 4)) we can write, for example, the second relation (2. 5) in the form

$$\nabla'_i (e'_m \overset{i}{e}) = \nabla'_m (e'_i \overset{i}{e}) \quad .$$

(2. 6a)

Further, because of ([2] , (1. 5)) the relation (2. 6a) be - comes

$$(\nabla'_i e'_m) \overset{i}{e} - (\nabla'_m e'_i) \overset{i}{e} = e'_i \frac{\partial x^k}{\partial x'^m} \nabla_k \overset{i}{e} - e'_m \frac{\partial x^k}{\partial x'^i} \nabla_k \overset{i}{e} \ .$$

(2. 6b)

We now apply ([2] , (1. 3c)) and (1. 9) to deduce the relation (2. 3) from (2. 6b).

In the case $m = n$ the problem of interrelation of coefficients of connection with the fundamental tensor ([2] , Chapter II) can also be discussed from the standpoint that the coefficients of connection can be represented in the form (2. 1). We point out that all the three indices run from 1 to n and that each set of coefficients of connection contains n^3 coefficients or $2n^3$ real and imaginary parts. Since in Chapter I we restricted the discussion to case A , we have to discuss in detail cases $Aa)$ and $Ab)$ ([1] , Chapter III).

Case Aa)

We know from previous results ([2] , Chapter II , Case Aa)) that for $m = n$ we have $n^2 \cdot (n-1)$ parts of coefficients of connection $(\Gamma_{\dot{j}})_{ki}$ at our disposal. The most simple requirement would therefore be that the antisymmetric parts, or what is the same, the torsion tensor should vanish

$$(\Gamma_{\dot{j}})^{A}_{ki} = 0 \,, \qquad {}^{>}S^{<<} = 0 \,. \tag{2.7}$$

Under the assumption (2.7), the relation ([2] , (2.11)) takes the form

$$\partial_k \, {}_i g_j = Re\,(\Gamma_{\dot{j}})^{S}_{ki} + Re\,(\Gamma_{\dot{i}})^{S}_{kj} + i\,[\,Im\,(\Gamma_{\dot{j}})^{S}_{ki} + Im\,(\Gamma_{\dot{i}})^{S}_{kj}\,] \tag{2.8}$$

After the substitution of the expression ([1] , (3.16)) for the fundamental tensor it is possible to separate the real and imaginary parts of (2.8). Thus, for the real parts, we obtain

$$\partial_k \, {}_i g_j^{1S} = Re\,(\Gamma_{\dot{j}})^{S}_{ki} + Re\,(\Gamma_{\dot{i}})^{S}_{kj} \quad ,$$

$$\partial_j \, {}_{i}g_k^{1s} = Re\,(\Gamma_k)_{ji}^{s} + Re\,(\Gamma_i)_{jk}^{s} \;,$$

$$\partial_i \, {}_{k}g_j^{1s} = Re\,(\Gamma_j)_{ik}^{s} + Re\,(\Gamma_k)_{ij}^{s} \;, \qquad (2.9)$$

where the last two equations follow from the first equation by an adequate change of indices. By subtracting the third equation from the sum of the first two expressions one determines the real part of the coefficient of connection

$$Re\,(\Gamma_i)_{kj}^{s} = \frac{1}{2}\left[\partial_k \, {}_{i}g_j^{1s} + \partial_j \, {}_{i}g_k^{1s} - \partial_i \, {}_{k}g_j^{1s}\right] = \left\{ {}_{k}{}^{i}{}_{j} \right\}^{1s} . \qquad (2.10a)$$

The imaginary part follows in an analogous manner

$$Im\,(\Gamma_i)_{kj}^{s} = \left\{ {}_{k}{}^{i}{}_{j} \right\}^{2s} . \qquad (2.10b)$$

The coefficients of connection are therefore completely determined in terms of the symmetric fundamental tensor starting with

$$(\Gamma_i)_{kj}^{s} = \left\{ {}_{k}{}^{i}{}_{j} \right\}^{s} = \left\{ {}_{k}{}^{i}{}_{j} \right\}^{1s} + i\left\{ {}_{k}{}^{i}{}_{j} \right\}^{2s} \qquad (2.11)$$

and (2.7).

Case Ab)

From previous results ([2] , Chapter II , Case Ab)) we know that in the case $m = n$ we have at our disposal $n^3 = n^2(n-1)/2 + n^2(n+1)/2$ parts of the coefficients of connection $(\Gamma_{\dot{j}})_{ki}$. The most simple requirement would therefore be that the antisymmetric real parts and the symmetric imaginary parts should vanish

$$Re\,(\Gamma^{\cdot}_{\dot{j}})^{A}_{kl} = 0 \,, \qquad Im\,(\Gamma^{\cdot}_{\dot{j}})^{S}_{ki} = 0 \,. \tag{2.12}$$

Under the assumptions (2.12) and with the expression ([1] , (3.17)) for the fundamental tensor the relation ([2] , (2.11) takes the form

$$\partial_k \; {}_{i}q^{1S}_{j} + i\,\partial_k \; {}_{i}q^{2A}_{j} = Re\,(\Gamma^{\cdot}_{\dot{j}})^{S}_{ki} + Re\,(\Gamma^{\cdot}_{i})^{S}_{kj} + i\,[Im\,(\Gamma^{\cdot}_{\dot{j}})^{A}_{ki} - Im\,(\Gamma^{\cdot}_{i})^{A}_{kj}] \tag{2.13}$$

Again, we separate the real and imaginary parts to obtain the system (2.9) from (2.13) for real parts. For the imaginary parts the system slightly differs from (2.9), the plus sign on the right-hand side being changed into the minus sign and q^{1S} being substituted by q^{2A} . Finally, we obtain the relation

$$(\Gamma_i)_{kj} = Re\,(\Gamma_i)^S_{kj} + iIm\,(\Gamma_i)^A_{kj} = \left\{ {i \atop k\;j} \right\}^{1S} + i \left\{ {j \atop k\;i} \right\}^{2A} . \qquad (2.14)$$

The coefficients of connection are therefore completely determined in terms of the fundamental tensor starting with (2. 12) and (2. 14).

Real case A)

If we confine ourselves to real quantities the difference between cases Aa) and Ab) disappears. The fundamental tensor has only real symmetric components. The coefficients of connection are also real quantities and we write ([2] , (2. 11)) in the form

$$2\,(\Gamma^S_j)_{ki} = \partial_k\;{}_i g^S_j = (\Gamma_j)^S_{ki} + (\Gamma_i)^S_{kj} + (\Gamma_j)^A_{ki} + (\Gamma_i)^A_{kj} . \qquad (2.15)$$

For the real case $m = n$ we have at our disposal $n^2(n-1)/2$ parts in the real coefficients of connection $(\Gamma_j)_{ki}$. In general, we can therefore require $n^2(n-1)/2$

additional relations to (2. 15) to be fulfilled. These relations require that the antisymmetric parts of the coefficients of connection $(\Gamma^A_{j})_{ki}$ be determined in the following way, with the help of a given real antisymmetric tensor $_{i}\overset{A}{g}_{j}$

$$2\,(\Gamma^A_{j})_{ki} = \partial_k\ {}_{i}\overset{A}{g}_{j} = (\Gamma_{j})_{ki} - (\Gamma_{i})_{kj} =$$

$$= (\Gamma_{j})^S_{ki} - (\Gamma_{i})^S_{kj} + (\Gamma_{j})^A_{ki} - (\Gamma_{i})^A_{kj}\ . \tag{2.16}$$

From (2. 15) and (2. 16) we deduce the relations

$$2\,(\Gamma_{j})_{ki} = \partial_k\,({}_{i}\overset{S}{g}_{j} + {}_{i}\overset{A}{g}_{j}) = \partial_k\ {}_{i}\overset{+}{g}_{j}\ ,$$

$${}_{i}\overset{+}{g}_{j} = {}_{j}\bar{g}_{i}\ , \tag{2.17}$$

$$2\,(\Gamma_{i})_{kj} = \partial_k\,({}_{i}\overset{S}{g}_{j} - {}_{i}\overset{A}{g}_{j}) = \partial_k\ {}_{i}\bar{g}_{j}\ .$$

By the same procedure as in cases $Aa)$ and $Ab)$ we now obtain from (2. 15) and (2. 16) the following relations for the symmetric and antisymmetric parts (in outer indices) of the coefficients of connection $(\Gamma_{i})_{kj}$

$$(\Gamma_{i})^S_{kj} + (\Gamma_{j})^A_{ki} + (\Gamma_{k})^A_{ji} = \left\{{}_{k}{}^{i}{}_{j}\right\}^S , \tag{2.18}$$

$$(\Gamma_k)^S_{ij} - (\Gamma_i)^S_{jk} + (\Gamma_j)^A_{ki} = \left\{ {}_{k}{}^{i}{}_{j} \right\}^A .$$

For the symmetric and antisymmetric parts (in outer indices) we find the following expressions from (2. 17) and (2. 18)

$$2\,(\Gamma_j)^S_{ki} = \left\{ {}_{k}{}^{j}{}_{i} \right\}^- + (\Gamma_k)_{ji} = \frac{1}{2}\left[\partial_k\, {}_i\overset{+}{g}_j + \partial_i\, {}_k\overset{+}{g}_j \right] ,$$

$$2\,(\Gamma_j)^A_{ki} = \left\{ {}_{k}{}^{i}{}_{j} \right\}^+ - (\Gamma_k)_{ji} = \frac{1}{2}\left[\partial_k\, {}_i\overset{+}{g}_j - \partial_i\, {}_k\overset{+}{g}_j \right] . \qquad (2.\,19)$$

Thus, the coefficients of connection $(\Gamma_j)_{ki}$ i. e. all their symmetric and antisymmetric parts (2. 15), (2. 16) and (2. 19) are completely determined with the help of the real symmetric fundamental tensor and the real antisymmetric tensor ${}_i\overset{A}{g}_j$.

For the particular case of real symmetric coefficients of connection $(\Gamma_j)^S_{ki}$ their antisymmetric part in outer indices (2. 19) must vanish. In this case we obtain the condition

$$(\Gamma_j)^A_{ki} = 0 , \qquad \partial_k\, {}_i\overset{+}{g}_j = \partial_i\, {}_k\overset{+}{g}_j . \qquad (2.20)$$

As a consequence from (2. 20), (2. 17) and (2. 18) we obtain the relation

$$(\Gamma_j)^{S}_{ki} = \tfrac{1}{2}\,\partial_k \, {}_i\overset{+}{g}_j = \left\{ {}_k{}^j{}_i \right\}^S . \qquad (2.21)$$

Further, the relation (2. 16) takes this form

$$2\,(\Gamma^A_j)_{ki} = \partial_k \, {}_i g^A_j = \left\{ {}_k{}^j{}_i \right\}^S - \left\{ {}_k{}^i{}_j \right\}^S = \partial_i \, {}_j g^S_k - \partial_j \, {}_i g^S_k \qquad (2.22)$$

which at the same time expresses the tensor ${}_i g^A_j$ in terms of the fundamental tensor ${}_i g^S_j$. The symmetric and antisymmetric parts (2. 15) and (2. 22) verify (2. 21) again.

At the end we recall the operator relation ([2] , (3. 13))

$$[\Theta] \equiv d_2 x^k \, d_1 x^j \, [\Theta]_{jk} + (\Theta_1 (d_2 x^k) - \Theta_2 (d_1 x^k))\, \Theta_k \qquad (2.23)$$

and determine its explicit form for $\Theta \equiv D, \delta$, for the metric space, what we were not able to do in ([2] , Chapter III). In the metric space the coordinate differentials are contravariant components of the displacement vector (1. 6). The last term on the right-hand side can therefore be determined with the help of ([2] , (1. 7), (1. 13)), e. g.

$$D_1 (d_2 x^k) = d_1 d_2 x^k + (\Gamma^k)_{ir} d_2 x^r d_1 x^i = d_1 d_2 x^k - \delta_1 (d_2 x^k) . \tag{2.24}$$

With the results (2.24) and (2.4) we write the operator relation (2.23) for the covariant and parallel displacement differentials in the final forms

$$[D] = d_2 x^k d_1 x^i \{ [D]_{ik} + {}^r S_{ik} D_r \} ,$$

$$[\delta] = d_2 x^k d_1 x^i \{ [\delta]_{ik} + {}^r S_{ki} \delta_r \} . \tag{2.25}$$

For the particular case of the symmetric coefficients of connection (2.20) the torsion tensor (2.4) disappears. In this case, therefore, the relations (2.25) between the commutator of the covariant (parallel displacement) differentials and derivatives have the same structure as for the absolute (ordinary) differentials ([2] , (3.14), (3.15)) in the metric space, too.

CHAPTER III

The curvature of the metric space

In a previous paper [2] we studied the second-order difference between the parallelly displaced quantities and determined it with the help of the commutator of the absolute differentials corresponding to two possible parallel displacements from $P(x^k)$ to $Q(x^k + d_1 x^k + d_2 x^k)$.

Further, the mentioned difference was expressed with the help of the quantity R or the second-rank tensor $_{>}P^{<}$ ([2] , (3.19), (3.23)). In the metric space this difference ([2] , (3.5)) expresses an intrinsic property of the space which we call the curvature of the space. In the metric space the quantity R becomes a fourth-rank tensor, as seen from its transformation property ([2] , (3.25)) and (1.9)

$$\begin{aligned} {}^{\bar{r}}P'_{\bar{i}} &= {}^{\bar{r}}R'_{\bar{k}\bar{j}\bar{i}}\, d_1 x'^{\bar{j}} d_2 x'^{\bar{k}} = \\ &= {}^{r}R_{kji} \frac{\partial x'^{\bar{r}}}{\partial x^{r}} \frac{\partial x^{k}}{\partial x'^{\bar{k}}} \frac{\partial x^{j}}{\partial x'^{\bar{j}}} \frac{\partial x^{i}}{\partial x'^{\bar{i}}}\, d_1 x'^{\bar{j}} d_2 x'^{\bar{k}} . \end{aligned} \tag{3.1a}$$

We therefore write the quantity R in the form

$$_{>}R^{<<<} = {}_{r}e\; {}^{r}R_{kji}\, e^{k} e^{j} e^{i} . \tag{3.1b}$$

and call it the curvature tensor.

Consequently, the formulas where the commutator of absolute differentials appears, can be written in terms of the curvature tensor, e.g. ([2] , (3.19))

$$[\Delta]\,\overset{i}{e} = {}_{>}R^{\underline{<2}\ \underline{<1}}{}_{<}\overset{i}{e}\ ,\quad [\Delta]\,e^{i} = e^{i}{}_{>}{}^{\underline{1>}\ \underline{2>}}R^{<}\ ,$$

$$[\Delta]\,\underset{i}{e} = {}_{>}R^{\underline{<2}\ \underline{<1}\ <}\underset{i}{e}\ ,\quad [\Delta]\,e_{i} = e_{i}{}^{>\ \underline{1>}\ \underline{2>}}R_{<}\ ,\qquad (3.2)$$

where the abbreviations have been used for the scalar products

$$\underline{<1} = \left({}^{<}{}_{>}(d_1 x)\right)\ ,\quad \underline{<2} = \left({}^{<}{}_{>}(d_2 x)\right)\ .$$

It is easy to prove that the commutator of absolute differentials of a vector (tensor) is a vector (tensor) with components equal to the commutator of the covariant differentials of the corresponding components of the original vector (tensor). Thus, we have from ([2] , (1.7), (1.8)) and (2.25), e.g.

$$[\Delta]^{>}a = {}^{i}e[D]\,{}_{;i}a\,,\quad [\nabla]_{jk}{}^{>}a = {}^{i}e\left\{[D]_{jk} + {}^{r}S_{jk}\,D_{r}\right\}{}_{;i}a\,,$$

$$[{}^{>}_{>}T^{<}{}_{<<}\,] = {}^{>}{}_{>}([D]T)^{<}{}_{<<}\,, \tag{3.3}$$

$$[\nabla]_{qs}({}^{>}_{>}T^{<}{}_{<<}\,) = {}^{j}e\,{}_{k}e\left\{[D]_{qs} + {}^{t}S_{qs}\,D_{t}\right\}{}_{j}{}^{k}T_{\ell}{}^{pr}\,e^{\ell}\,e_{p}\,e_{r}\,.$$

We find the relation of covariant differentials (derivatives) and the curvature tensor by comparing the expressions(3. 3) with the relations ([2] , (3. 26), (3. 27)) or by directly substituting (3. 2) into the left-hand side of (3. 3).

Two basic properties of the components of the curvature tensor with respect to the indices k and j were determined in ([2] , (3. 20), (3. 21))

$$\begin{aligned} {}_{r}R_{kj}{}^{i} &= -\,{}_{r}R_{jk}{}^{i}\,, \\ {}_{r}R_{kj}{}^{i} &= {}_{rkj}R^{i}\,. \end{aligned} \tag{3.4}$$

We now try to find other basic properties and relations for the curvature tensor for the real case A . First, we closely examine the absolute derivatives and their commutators of the basis vectors. Because of ([2] (1.3c),(2. 17)) and (2. 4) for the contravariant tangent basis vectors we immediately find the relations

$$\nabla_k \, {}_i e - \nabla_i \, {}_k e = {}_p e \, {}^p S_{ki} \, , \quad \nabla_k \, e_i - \nabla_i \, e_k = {}^p S_{ki} \, e_p \, . \tag{3.5}$$

It follows from (3. 5) that the commutator of the absolute derivatives of basis vectors has the form

$$[\nabla]_{jk} \, {}_i e = \nabla_j \, (\nabla_i \, {}_k e + {}_p e \, {}^p S_{ki}) - \nabla_k (\nabla_i \, {}_j e + {}_p e \, {}^p S_{ji}) \quad .$$

After easy transformations we obtain the final result

$$[\nabla]_{ij} \, {}_k e + [\nabla]_{jk} \, {}_i e + [\nabla]_{ki} \, {}_j e =$$

$$= \nabla_i \, ({}_p e \, {}^p S_{jk}) + \nabla_j \, ({}_p e \, {}^p S_{ki}) + \nabla_k \, ({}_p e \, {}^p S_{ij}) \; = {}_{>} (ijk) \; . \tag{3.6a}$$

By a completely identical procedure for the bra contravariant basis vectors an analogous relation is obtained

$$[\nabla]_{ij} \, e_k + [\nabla]_{jk} \, e_i + [\nabla]_{ki} \, e_j = (ijk)_{<} \; . \tag{3.6b}$$

Further, we recall two facts : First, the commutator of absolute derivatives of every member of the scheme ([1] , (1.6)) vanishes, e.g.

$$[\nabla]_{jk}\ {}_r g_i = [\nabla]_{jk}\ (e_r\ {}_i e) = ([\nabla]_{jk}\ e_r)\ {}_i e + e_r([\nabla]_{jk}\ {}_i e) = 0\ . \tag{3.7}$$

Second, the scalar product is commutative ([1] , (3.11))in the real case A . From (3.6b) and (3.7) we therefore deduce the expression

$$([\nabla]_{jk}\ e_i)\ {}_r e = -\, e_k([\nabla]_{ji}\ {}_r e) + e_j\cdot([\nabla]_{ki}\ {}_r e) + (jki)_<\ {}_r e\ .$$

We now substitute the respective expressions (3.6) for the commutators of absolute derivatives, and applying the commutativity of scalar products we obtain the final result in the form

$$\begin{aligned} &2([\nabla]_{jk}\ e_i)\ {}_r e + e_k\ {}_>(jir) - e_j\ {}_>(kir) = \\ &= 2([\nabla]_{ir}\ e_j)\ {}_k e + (jki)_<\ {}_r e - (jkr)_<\ {}_i e\ . \end{aligned} \tag{3.8a}$$

An analogous relation is obtained by a completely identical procedure for the ket contravariant basis vectors

$$2e_r([\nabla]_{jk}\ {}_i e) + (jir)_<\ {}_k e - (kir)_<\ {}_j e = \tag{3.8b}$$

$$= 2 e_k ([\nabla]_{ir} \; {}_j e) + e_r \, {}_>(jki) - e_i \, {}_>(jkr) \; .$$

Substituting the commutators of the absolute derivatives of the basis vectors ([2] , (3. 19)) or (3. 2) into the relations (3. 6), (3. 7) and (3. 8) we obtain the relations for the components of the curvature tensor. From (3. 6) two identical formulas are derived

$$^{q}R_{kjr} + {}^{q}R_{rkj} + {}^{q}R_{jrk} = e^{q} \, {}_>(jkr) \, ,$$

$$(3.9)$$

$$_{rjk}R^{q} + {}_{jkr}R^{q} + {}_{krj}R^{q} = (jkr)_< \, {}^{q}e \; .$$

The relation (3. 7) gives the relation (3. 4_2) again. The relations (3. 8) yield two identical formulas

$$2 \; {}_rR_{kji} + (jir)_< \, {}_ke - (kir)_< \, {}_je = 2 \; {}_jR_{rik} + e_r \, {}_>(jki) - e_i \, {}_>(jkr) \, ,$$

$$(3.10)$$

$$2 \; {}_{ijk}R_r + e_k \, {}_>(jir) - e_j \, {}_>(kir) = 2 \; {}_{kir}R_j + (jki)_< \, {}_re - (jkr)_< \, {}_ie \; .$$

The abbreviations (3. 6a) ${}_>(jkr)$ in the relations (3. 9) and (3. 10) have to be substituted by the following explicit expression

$$ {}_{>}(jkr) = {}_{p}e \; S \left\{ \partial_j \; {}^{p}S_{kr} + (\Gamma^{p})_{jq} \; {}^{q}S_{kr} \right\} , \tag{3.11} $$

where the symbol S means the cyclic sum over the indices j, k and r. The expression for $(jkr)_{<}$ is identical with (3.11), ${}_{p}e$ being changed into e_p.

When the torsion tensor (2.4) and consequently the expression (3.11) vanish, the relations (3.9) and (3.10) can greatly be simplified. Thus,for the particular case of real symmetric coefficients of connection (2.20)from (3.9) we obtain the relation

$$ {}^{q}R_{kjr} + {}^{q}R_{rkj} + {}^{q}R_{jrk} = 0 \tag{3.12a} $$

and by multiplying it with ${}_{i}g_{q}$ we have

$$ {}_{i}R_{kjr} + {}_{i}R_{rkj} + {}_{i}R_{jrk} = 0 . \tag{3.12b} $$

From (3.10) the relation between the components of the curvature tensor is obtained

$$ {}_{r}R_{kji} = {}_{k}R_{rij} . \tag{3.13} $$

We proved in ([2], (3.7),(3.8), (3.10), (3.11)) that for every differential operator Θ, for which the product rule is valid, the product rule for the commutator of such two operators is valid, too

$$[\Theta]_{12}(A\cdot C) = ([\Theta]_{12}A)\cdot C + A\cdot([\Theta]_{12}C) \ . \tag{3.14}$$

It is easy to prove that for such three operators Θ_1, Θ_2 and Θ_3 the following identity holds

$$\{\Theta\} = [\Theta_1[\Theta]_{23}] + [\Theta_2[\Theta]_{31}] + [\Theta_3[\Theta]_{12}] = 0 \ . \tag{3.15}$$

Moreover, for the operators formed with Θ operators in the following manner

$$K_1 = [\Theta]_{23} \ , \quad K_2 = [\Theta]_{31} \ , \quad K_3 = [\Theta]_{12} \ , \tag{3.16a}$$

$$L_1 = [K]_{23} \ , \quad L_2 = [K]_{31} \ , \quad L_3 = [K]_{12} \ , \tag{3.16b}$$

----

the identities analogous to (3.15) are valid

$$\{K\} = [K_1[K]_{23}] + [K_2[K]_{31}] + [K_3[K]_{12}] = 0 , \tag{3.17a}$$

$$\{L\} = [L_1[L]_{23}] + [L_2[L]_{31}] + [L_3[L]_{12}] = 0 , \tag{3.17b}$$

In order to derive some more relations involving the components of the curvature tensor, we first discuss the operator identities obtained for the case of absolute differentials $\Theta \equiv \Delta$. We start with

$$\{\Delta\}\, a_{<} = 0 \ . \qquad (3.18)$$

In order to determine the explicit form of the left-hand side (3.18), we need the expressions ([2] , Chapter I and Chapter III)

$$\Delta_1 a_{<} = (D_1\ a^i)\, e_i \quad ,$$

$$\Delta_3 [\Delta]_{12}\, a_{<} = D_3 (a^i [\Delta]_{12}\, e_i) \quad , \qquad (3.19)$$

$$[\Delta]_{12}\ \Delta_3\, a_{<} = (D_3\ a^i)([\Delta]_{12}\, e_i) \ ,$$

and two more expressions for each of them by cyclic change of indices 1, 2 and 3. Substituting the expressions (3.19) into the explicit form (3.15) of (3.18), we conclude that the identical vanishing of (3.18) leads to the relation

$$D_1 (K_1\, e_i) + D_2 (K_2\, e_i) + D_3 (K_3\, e_i) = 0 \ . \qquad (3.20)$$

Three more relations of the form (3.20) for the basis vectors e^i , ${}_i e$ and ${}^i e$ could be deduced in a similar manner.

The identity (3. 17a) can be analyzed in an analogous way

$$\{K\}\, a_4 = 0 \quad . \tag{3.21}$$

Here we need the relations

$$K_1\, a_4 = (D_2 D_3 - D_3 D_2)\, a^i e_i = [D]_{23}\, a^i e_i \quad ,$$

$$K_3 [K]_{12}\, a_4 = [D]_{12}\, (a^i [K]_{12}\, e_i) \quad , \tag{3.22}$$

$$[K]_{12} K_3\, a_4 = ([D]_{12}\, a^i)[K]_{12}\, e_i \quad ,$$

and two more relations for each of them by cyclic change of indices 1, 2 and 3. Substituting the expressions (3. 22) into the explicit form (3. 15) of (3. 22), the identical vanishing of (3. 22) leads to the relation

$$[D]_{12}\, (L_3 e_i) + [D]_{23}\, (L_1 e_i) + [D]_{31}\, (L_2 e_i) = 0 \, . \tag{3.23}$$

Three more relations of the form (3. 23) for the basis vectors e^i , ${}_i e$ and ${}^i e$ could be deduced in a similar manner.

In this way we could proceed with building more and more complicated expressions which contain the covariant differentials and the absolute differentials of the basis vectors. Instead of doing so, we illustrate by a simple example of (3. 20) that these relations represent expressions involving the components of the curvature tensor. Here we have to substitute the expressions (3.2) for the commutators of absolute differentials (3. 16a) to obtain from(3. 20) the relation

$$D_1(d_2 x^k d_3 x^m \; {}_{ikm}R^r) + D_2(d_3 x^m d_1 x^j \; {}_{imj}R^r) +$$

$$+ D_3(d_1 x^j d_2 x^k \; {}_{ijk}R^r) = 0 \; .$$

We now apply the product rule for the covariant derivatives ([2] , (1. 16)), recall that the coordinate differentials are contravariant components of the displacement vector (1. 6), and consequently their covariant derivatives are given by ([2] , (1. 7)). Finally, after an easy treatment we obtain (3. 20) with the help of (3. 4) in the form

$$\begin{aligned} &D_j \, ({}_{ikm}R^r) + D_k({}_{imj}R^r) + D_m \, ({}_{ijk}R^r) + \\ &+ {}^qS_{jk} \, {}_{imq}R^r + {}^qS_{km} \, {}_{ijq}R^r + {}^qS_{mj} \, {}_{ikq}R^r = 0 \; . \end{aligned} \tag{3. 24}$$

For the particular case of real symmetric coefficients of connection (2. 20) the last three terms in (3.24) vanish due to the disappearance of the torsion tensor. Thus, in this case the following relation exists for the covariant derivatives of the components of the curvature tensor

$$D_j \left({}_{ikm}R^r \right) + D_k \left({}_{imj}R^r \right) + D_m \left({}_{ijk}R^r \right) = 0 \quad . \qquad (3.25)$$

Three relations similar to (3.24) and (3.25) follow from (3.20), written for the basis vectors e^i , ${}_ie$ and ${}^ie$.

We are not going here into further details of the theory. However, to emphasize the essential features of the described general theory and some important particular cases (space with affine connection, metric space, Riemann space etc.), we add some supplementary remarks.

CHAPTER IV

Remarks

a) The basic notions

The formulation of the vector and tensor calculus given in the first part of this paper [2] was concerned with a more general case of not necessarily equal dimensions of the parameter manifold and vector spaces $X(P)$, $P \in \Omega$. In the present part of the paper a very important particular case of equal dimensions, $m = n$, is considered. Such an approach is suitable to find out the features of the calculus derived from the properties of the manifold and the vector space itself.

The whole calculus is based on the fundamental notion of the absolute differential (derivative) $\Delta (\nabla_k)$ ([2] , Chapter I). General, compact and simple rules for the determination of the absolute differential (derivative) and for their commutators have been established (1.5), (3.10) and applied to scalars (1.4), (3.16), vectors (1.7), (3.26) and tensors (1.8), (3.27) in ([2] , Chapter I, III). Further, the connection of the absolute differential with the operators acting on scalars, vector and tensor components has been determined: ordinary differential (partial derivative) d (∂_k) , covariant differential (derivative) D (D_k) and parallel displacement differential (derivative) $\delta (\delta_k)$ ([2] , Chapter I). These symbols have been consistently used through the paper, but it should be mentioned that they may differ from those found elsewhere [3] .

For the particular case $m = n$ the

definition (Chapter I) of the displacement vector is of fundamental importance. With the help of this definition it was possible in case αA to identify the symmetric part of the fundamental tensor as the metric tensor (1.4) and to express the transformation coefficients in terms of the partial derivatives of the coordinates (1.9).

b) The fundamental tensor

The valence raising and lowering operators ([1] , (2.4)) were called the fundamental tensors. Their components are symmetric in case Aa), Hermitian in case Ab), orthogonal in case Ba) and unitary in case Bb)([1], Chapter III). For the real case $\alpha\, A\, a$ and the complex case $\alpha\, Ab$) the distance (1.2), (1.4) is determined with the help of the symmetric part of the fundamental tensor.

It is interesting to note that for the same metric tensor (i. e. the same real symmetric part of the fundamental tensor) the complex case $\alpha\, Ab$) compared with the real case $\alpha\, Aa$) contains $s = n(n-1)/2$ supplementary degrees of freedom (i. e. the components of the imaginary antisymmetric part of the fundamental tensor). In spite of the inconvenience to deal with complex quantities, this fact was used in Einstein's first attempt [4] to formulate the geometric frame for the unified theory in the four-dimensional space. In this case ($n = 4$) s equals six and corresponds to the number of components of the antisymmetric tensor of the electromagnetic field $F_{\lambda\mu}$

On the other hand in the frame of the described approach, in the real case αA we have the possibility of forming the basic tensor ${}_i\overset{+}{g}_j = {}_i\overset{S}{g}_j + {}_i\overset{A}{g}_j$ (2. 17), whose symmetric part is the fundamental (metric) tensor. The basic tensor completely determines the coefficients of connection (2. 19), the $s = n(n-1)/2$ supplementary degrees of freedom deriving from the antisymmetric tensor ${}_i\overset{A}{g}_j$ at our disposal. It could be hoped that the given approach with such a basic tensor ${}_i\overset{+}{g}_j$ could be useful for the formulation of the geometric frame of the unified theory. In this way some ambiguities and arbitrariness could be avoided in the existing formulations with the nonsymmetric fundamental or metric tensor [5].

In the particular case ([1], Chapter II)

$$ {}_ig_j(P) = {}_i\delta_j \qquad {}^{ij}g(P) = {}^{ij}\delta , \quad P \in \Omega \tag{4. 1}$$

the covariant and contravariant forms of vectors and tensors are identical. The basis vectors are orthonormal and in the real case A they become unit vectors

$$e_i \; {}_i e = {}_i\delta_i \;, \quad \sqrt{e_i \; {}_i e} = \|e_i\| = 1 \;. \tag{4.2}$$

In the metric space with (4.1) the square of distance is positive definite (Euclidean space) in the Cartesian coordinates x^k, for which the described basis vectors (4.2) are tangent basis vectors

$$ds^2 = dx^i \, dx^i > 0 \;. \tag{4.3}$$

We now take another system of coordinate lines ([2] , (1)) in the Euclidean space, the so-called generalized coordinates. In the generalized coordinates x'^i the distance (4.3) will be expressed in the form

$$ds^2 = \frac{\partial x^i}{\partial x'^k} \frac{\partial x^i}{\partial x'^p} \, dx'^k \, dx'^p = {}_k g'_p \, dx'^k dx'^p \,. \tag{4.4}$$

Hence we see that the components of the metric tensor are symmetric in the system of generalized coordinates. The basis vectors, tangent to the generalized coordinate lines ([1] , Chapter IV) (1.9), are

$$e'_k = (e'_k\, {}_r e)\, e_r = \frac{\partial x^r}{\partial x'^k}\, e_r \;, \quad e'^k = (e'^k\, {}_r e)\, e_r = \frac{\partial x'^k}{\partial x^r}\, e_r \;, \tag{4.5}$$

$${}_k e' = {}_r e\,(e_r\, {}_k e') = {}_r e\, \frac{\partial x^r}{\partial x'^k} \;, \quad {}^k e' = {}_r e\,(e_r\, {}^k e') = {}_r e\, \frac{\partial x'^k}{\partial x^r} \;.$$

The fundamental scheme ([1] , (1.6)) for these basis vectors takes the form

$$e'^i\, {}_j e' = {}^i\delta_j \;, \qquad e'_i\, {}^j e' = {}_i\delta^j \;, \tag{4.6}$$

$$e'^i\, {}^j e' = {}^i g'^j = \frac{\partial x'^i}{\partial x^r}\,\frac{\partial x'^j}{\partial x^r} \;, \quad e'_i\, {}_j e' = {}_i g'_j = \frac{\partial x^r}{\partial x'^i}\,\frac{\partial x^r}{\partial x'^j} \;.$$

From (4. 6) the symmetry of the fundamental tensor components is evident, in accordance with (4. 4).

For the particular case of linear coordinate transformations ($a^i_{\ k}$, a^i constants)

$$x'^j = a^j_{\ k}\, x^k + a^j \;, \quad x^k = A^k_{\ j}\, x'^j + A^k \;, \quad \det\,(a^j_{\ k}) \neq 0 \;;$$

$$a^j_{\ k}\, A^k_{\ i} = \delta^j_{\ i} \;, \quad A^k_{\ j}\, a^j_{\ i} = \delta^k_{\ i} \;, \tag{4.7}$$

$$A^k_{\ j}\, a^j = -A^k \;, \quad a^j_{\ k}\, A^k = -a^j \;,$$

we obtain the constant components of the symmetric fundamental tensor in the system of generalized coordinates x'^i

$$g'^{ij} = a^i_{\ k}\, a^j_{\ k}\ , \quad g'_{ij} = A^r_{\ i}\, A^r_{\ j}\ .$$

Moreover, we see from (4. 7) that the difference of the Cartesian coordinates of every pair of points $P\ ,\ Q \in \Omega$ transforms as vector components, the transformation coefficients being

$$\frac{\partial x'^i}{\partial x^k} = a^i_{\ k}\ , \quad \frac{\partial x^k}{\partial x'^i} = A^k_{\ i}\ . \tag{4. 8}$$

Thus, in this case it is possible to introduce the "radius-vector" with the components in the Cartesian system

$$\vec{r} = \overrightarrow{PQ} = (x^i_Q - x^i_P)\, e_i\ . \tag{4. 9a}$$

The norm of the radius vector

$$r = \overline{PQ} = \sqrt{\sum_i (x^i_Q - x^i_P)^2} \tag{4. 9b}$$

is identical with the distance (1. 4) between the points P and Q calculated along the straight line $x^i = (1-t)\, x^i_P + t\, x^i_Q$, $i = 1, \ldots, n$; determined by the points $P\,(t=0)$ and $Q\,(t=1)$.

c) The coefficients of connection

The coefficients of connection were introduced into the theory by means of the absolute differentials of the basis vectors ([2] , (1.3c)). The basic relations for them ([2] , (2. 10), (2. 11)) were obtained by determining the absolute differentials of the quantities occurring in the fundamental scheme ([1] , (1. 6)). An important consequence was the vanishing of the absolute differential (derivative) of the fundamental tensor ([2] , (2. 12)). The basic relations connect the coefficients of connection with the components of the fundamental tensor. Also, they give the possibility for a detailed study and analysis of these coefficients, especially for the determination of their symmetric and antisymmetric parts with the help of the fundamental and basic tensor ([2] , Chapter II), (Chapter III).

The transformation law for the coefficients of connection ([2] , (2. 7), (2. 8)), (2. 2) showed that they are not tensors. However, for the linear transformation (4. 7) they transform as components of a third-rank tensor. In this case the second term of the right-hand side in (2. 2) vanishes identically.

In a metric space with real symmetric (in outer indices) coefficients of connection $(\Gamma_i)^{s}_{jk}$, i. e. in a Riemann space, the coefficients of connection are completely determined by the real metric tensor in terms of the Riemann-Christoffel symbols (2. 10).

In the particular case of a Riemann space with the metric tensor (4. 1) the symmetric coefficients of connection disappear identically in the Cartesian coordinates and in all rectilinear coordinate systems obtained from these coordinates by the linear transformation (4. 7). From ([2] , (1. 3c)) we infer that the absolute differentials of the basis vectors vanish identically for the rectilinear coordinates. Thus, we have $e\,(P)_a \equiv e\,(Q)$ for all $P,\ Q \in \Omega$, i. e. the parallelly displaced tangent basis vectors $e\,(P)_a$ are tangent basis vectors $e\,(Q)$ at the point Q , the term "rectilinear" being thus explained.

In the real case αA from both relations (2. 2) we obtain the expression for the coefficients of connection in generalized coordinates

$$(\Gamma'^{i})_{r\ell} = \frac{\partial x'^{i}}{\partial x^{p}} \frac{\partial}{\partial x'^{r}} \left(\frac{\partial x^{p}}{\partial x'^{\ell}} \right) \quad . \qquad (4.10)$$

d) The curvature

In the first part of this paper ([2], Chapter III) we discussed the difference between the parallelly displaced quantities for two different paths $P \to Q_1 \to Q$ and $P \to Q_2 \to Q$ and expressed it with the help of the commutator of the absolute differentials. We found the explicit expressions for the commutator of the absolute differentials of the basis vectors ([2], (3. 19)) given in abbreviated forms ([2], (3. 23)) in terms of the quantity R or of the linear second-order operator $_{>}P^{<}$. Then, it was easy to determine the commutator of absolute differentials applied to scalars ([2], (3. 16)), vectors ([2], (3. 26)) and tensors ([2], (3. 27)).

In the case $m = n$ the quantity R becomes the fourth-rank tensor, the so-called curvature tensor (3. 1). The fundamental properties of the curvature tensor are expressed by the relations (3. 4), (3. 9), (3. 10) and (3. 24). The particular forms of these relations in the Riemann space (3. 12), (3. 13) and (3. 25) are easily deduced from the general case by requiring the torsion tensor to vanish. These results were obtained by an easy and direct procedure, essentially based on the absolute differentials of the basis vectors.

We have seen ([2] , Chapter III) that the vanishing of the commutator of absolute differentials is the expression of the absolute parallelism. Thus, in the case $m = n$ the vanishing of the curvature tensor is the expression of the absolute parallelism in a space. In the Euclidean space the curvature tensor vanishes identically because the coefficients of connection disappear identically in the Cartesian coordinates.

We could ask what conditions have to be fulfilled in order to find a coordinate system in which the coefficients of connection vanish. By assuming this coordinate system to be labelled by x^k, $\Gamma(x^k) = 0$, in another (holonomic) coordinate system ([2] , (1)) x'^i we have the relations (4. 10) for the transformation coefficients and coefficients of connection

$$\frac{\partial x^p}{\partial x'^i} (\Gamma'^i)_{r\ell} = \frac{\partial}{\partial x'^r} \left(\frac{\partial x^p}{\partial x'^\ell} \right) \tag{4.11}$$

From (4. 11) and (2. 5) we deduce the first condition in the form

$$\frac{\partial x^p}{\partial x'^i} \, {}^i S'_{qr} = 0 . \tag{4.12}$$

Further, the commutativity of the partial derivatives of the transformation coefficients themselves, $\frac{\partial x^p}{\partial x'^l}$, must exist. Thus, after equating the corresponding r, k and k, r second partial derivatives of the transformation coefficients, with the help of the curvature tensor definition ([2], (3.19)) we obtain from (4.11) the other condition

$$\frac{\partial x^p}{\partial x'^i} \, {}_q R'_{rk}{}^i = 0 . \tag{4.13}$$

The conditions (4.12) and (4.13) have to be fulfilled for every allowed transformation of coordinates ([2], (1)). This means that the torsion tensor and curvature tensor have to vanish identically.

Finally, we would point out that following the remark Chapter VIe) in [1], in the real case A), we can go over to the notation with unilateral indices. To obtain the coincidence with the notations used in the literature we have to apply the case $Aa)$ bra, ket, right side ([1] (6.18a)). For illustration it is sufficient to indicate a few examples

$$ {}^i g^j \to g^{ji} \; ; \; {}_i g_j \to g_{ji} \; ; \; {}^i S_{qr} \to S_{qr}{}^i \; ; \; {}^q R_{kjr} \to R_{kjr}{}^q . \tag{4.14}$$

All formulas can easily be transcribed in this notation.

References

[1] Z. Janković : A contribution to the vector and tensor algebra ;

[2] Z. Janković : A contribution to the vector and tensor analysis I ;

[3] J.A.Schouten : Ricci-Calculus, 2. ed., Berlin 1954 ; R.K. Raševskij : Rimanova geometrija i tenzornij analiz, Moskva 1967 ;

[4] A. Einstein : A generalization of the relativistic theory of gravitation, Ann. of Math. 46 (1945) 578 ;

[5] A. Einstein - E.G. Strauss: A generalization of the relativistic theory of gravitation II, Ann. of Math. 47 (1945) 731 ;

E. Schrödinger : Space - time structure, Cambridge 1950;

A. Einstein : The meaning of relativity, 4. ed., Princeton 1953 ;

M.A. Tonnelat : La théorie du champ unifié d'Einstein, Paris 1955 ;

A. Lichnerowicz : Théories relativistes de la gravitation et de l'éléctromagnetisme, Paris 1955;

B. Finzi : Relatività generale e teorie unitarie (Cinquant'anni di relatività 1905 - 1955), Firenze 1955;

V. Hlavaty : Geometry of Einstein's unified field theory, Groningen, 1957 ;

A. Crumeyrolle : Sur quelques interprétations physiques et théoriques des équations du champ unitaire d'Einstein-Schrödinger, Riv. Math. Univ. Parma, I 3 (1962) 331; II 9 (1964) 85 ;

R.N. Sen : On new theories of space in general relativity, Bull. Calc. Math. Soc. 56 (1964) 1 ;

M.A. Tonnelat : Les théories unitaires de l'électromagnetisme et de la gravitation, Paris 1965.

Contents

Printed in the United States
By Bookmasters